National 4 & 5
DESIGN & MANUFACTURE
COURSE NOTES

Jill Connolly

© 2019 Leckie & Leckie Ltd
Cover image © Vitra International AG
Cover design © ink-tank and associates

001/06012019

10 9 8 7 6 5 4 3 2 1

All rights reserved. No part of this publication may be reproduced, stored in a retrieval system, or transmitted in any form or by any means, electronic, mechanical, photocopying, recording or otherwise, without the prior written permission of the Publisher or a licence permitting restricted copying in the United Kingdom issued by the Copyright Licensing Agency Ltd., 90 Tottenham Court Road, London W1T 4LP.

The author asserts her moral right to be identified as the author of this work.

ISBN 9780008282196

Published by
Leckie & Leckie Ltd
An imprint of HarperCollins*Publishers*
Westerhill Road, Bishopbriggs, Glasgow, G64 2QT
T: 0844 576 8126 F: 0844 576 8131
leckieandleckie@harpercollins.co.uk
www.leckieandleckie.co.uk

Printed in the Italy by Grafica Veneta S.P.A.

A CIP Catalogue record for this book is available from the British Library.

Acknowledgements

We would like to thank the following for permission to reproduce photographs.

P9b Mark Newson Ltd and Magis; P9c Loft Furniture Ltd; P10a Herman Miller; P15a Graham Murdoch and Real Wood Studios; P20 Charlotte Tangye Design; P23b Steelcase Inc; P24a Steelcase Inc; P28b Adidas; P31 Estudio Mariscal; P37 Desu Design; P40 Suck UK; P41c Branca, Lisboa; P49a British Standards Institution; P49b International Organisation for Standardisation; P49c European Commission; P50 Carrera; P60 www.doiydesign.com; P67 Ikea; P75b gallery: lapas77 / Shutterstock.com, iPad: manaemedia / Shutterstock.com, concert: Christian Bertrand / Shutterstock.com; P77a and P77b James DysonFoundation; P101 Snug. Studio, Germany; P109 © Yellow Broom; P124a Forestry Commission Scotland; P134bDraper tools; P136a–P137f Draper tools; P150c Draper Tools; P152a–P152b Draper tools; P152c HPC Laser Ltd; P153a Umbra; P157b–P157d Draper Tools; P157e Bowers Group; P157f Draper Tools; P165a Draper Tools; P165c Draper Tools; P166h Draper tools; P166i HME Technology Ltd; P175a and P175b Bonnie Bling; P179c and P179d One Foot Taller; P214 Estudio Mariscal

All other images © Shutterstock.com, © Thinkstock.com or Author's own.

Introduction

About this book

This book will provide you with the knowledge and understanding for the **National 4 and 5 qualifications in Design and manufacture**: **Unit 1**, **Design** and **Unit 2, Materials and manufacturing**.

As well as covering the learning that will prepare you for the N4 and N5 unit assessments and the N5 Design and manufacture question paper, it also leads you through the stages of the course assignment, helping you to design a product that is suitable for manufacture.

Features

CHAPTER SUMMARIES

Each chapter starts with a summary of the learning it contains.

> **In this chapter you will learn about:**
> - design activities
> - the members of a design team and their roles
> - the term 'commercial products'
> - the difference between needs and wants.

ACTIVITIES

The activities throughout the book aim to help you to develop your skills to become a successful designer.

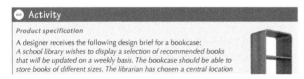

TEST YOUR KNOWLEDGE

These boxes contain questions to help you review your knowledge and understanding. Suggested answers can be found using the QR codes or by entering the web addresses into an internet browser.

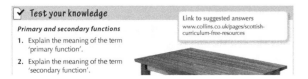

CASE STUDIES

These are real-life examples of relevant products and organisations, brilliant designers and exciting design projects.

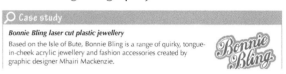

HINT

Hint boxes give extra support and make helpful suggestions for achieving success in your course.

MAKE THE LINK

Design and manufacture is not a subject in isolation! Make-the-link boxes provide appropriate links to other subject areas. You are also encouraged to 'Make the link' between different parts of the course, and between your knowledge and understanding and your practical skills.

CHECK YOUR PROGRESS

At the end of every chapter there is a checklist that enables you to rate your understanding of the key concepts, showing you where you need to target your efforts.

ASSESSMENT

Chapter 7 provides help and advice on the course assignment and written paper, along with example exam-style questions with suggested answers and marking commentary, and practice exam questions for you to try yourself.

Contents

Section 1 Design — 6

1 An introduction to designing — 8
Design — 8
Design activities — 10
The design team — 11
Commercial products — 16
Needs and wants — 17

2 The factors that influence design — 19
The factors that influence design — 19
Function — 20
Performance — 21
Market — 28
Methods to support sustainability — 34
Aesthetics — 35
Ergonomics — 42
Environmental considerations in design — 50

3 Designing — 54
The design brief — 54
Analysing a design brief — 56
Design brief example — 56
Research — 58
Carrying out research — 60
User trip — 61
Product specification — 71
Idea-generation techniques — 73
Design ideas — 78
Design proposal — 90
Sequence of operations — 90
Evaluation — 94

Section 2 Materials and manufacturing — 98

4 An introduction to materials — 100
Selecting a material — 100
Wood — 102
Metal — 110
Plastic — 114
Researching materials — 118

5 Manufacturing in the workshop — 126
Prepare for manufacture — 126
Purchasing materials and components — 126
Sequence of operations — 127
Manufacturing with wood — 133
Manufacturing with plastic — 150
Manufacturing with metal — 156
Metal threads — 159

6 Commercial manufacturing — 169
An introduction to commercial manufacture — 169
Computer aided manufacture — 170
3D printing — 172
Commercial wood and manufactured-board processes — 173
Commercial plastic processes — 173
Commercial metal processes — 180
Impact of design and manufacturing technologies — 182
Environmental impact of commercial manufacture — 183

Section 3 Assignment and exam preparation — 186

7 Assessment — 187
National 4 added value unit — 187
National 5 assessment — 190
National 5 course assignment — 190
National 5 question paper — 192
National 5 course specification for the question paper — 192
National 5 exam-style questions — 197
Practice exam-style questions — 207

CONTENTS

- **An introduction to designing**
- **The factors that influence design**
- **Evaluating products**
- **The process of designing**

DESIGN SECTION OVERVIEW

The first three chapters of this book will provide you with the knowledge and understanding for **Unit 1**, **Design**, and will also be useful when you complete your course assignment. Combined with your classwork and the guidance of your teacher, these chapters will lead you through the stages of the design process, helping you to design a product that is suitable for manufacture. The activities in this section aim to help you to develop your skills to become a successful designer. The information in this section will also be useful throughout the course, along with other reference books and additional sources of information.

Chapter 1, **An introduction to designing**, explains the process of designing and introduces you to the various members of a design team. It explains the term 'commercial products' and sets out the reasons why we have these products in our lives. This chapter provides a foundation on which to build your knowledge and understanding of designing.

Chapter 2, **The factors that influence design**, focuses on the design of commercial products and the way in which they are influenced by function, performance, market, aesthetics and ergonomics. In this chapter you will develop an understanding of these five main factors, which influence the design of commercial products.

The final chapter in this section is **Designing**. This chapter explains the tasks that designers work through, from identifying a need for a product, to reaching a final design proposal. There is a description of each stage of the design process, an exploration of the purpose of each stage and an elaboration of links between the stages. For those stages that involve the production of written work, such as the analysis and specification, helpful tips and suggestions are given to aid report structure and wording.

This chapter also explains the graphic and modelling techniques used to communicate and develop your ideas in 2D and 3D. Further support for the design development stage is provided, to help you to justify design decisions and apply relevant information regarding the design factors. The end of this section explains how to present the final design concept and make plans for the manufacture of your design.

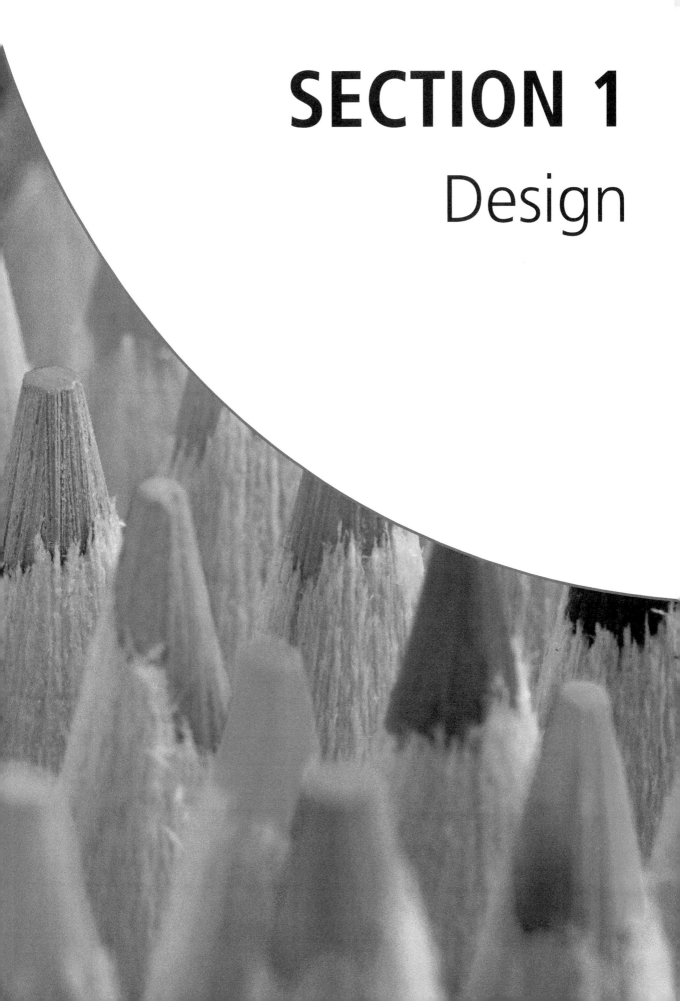

SECTION 1
Design

1 An introduction to designing

In this chapter you will learn about:
- design activities
- the members of a design team and their roles
- the term 'commercial products'
- the difference between needs and wants.

> **Hint**
> You can study a design movement to learn more about the impact of design and manufacturing technologies over a period of time, and compare that to modern products.

Design

Designing is about developing an idea, or **concept**, and then fine tuning the details so that the concept can be brought to life. Designing involves sketching, drawing and modelling the concept until what you create becomes true to the vision in your head. This involves making decisions and compromises. Designing includes selecting appropriate materials and manufacturing methods, and then figuring out how all the pieces can be put together to make something useful or beautiful ... or both.

Many famous designers have shifted the way in which society uses and values products. For example, the leader of the Arts and Crafts design movement in the 1890s was a man called William Morris. This was a time when everyday products were beautiful, hand-crafted items and he was a driving force in changing people's minds about having such things in their homes. He famously said: 'Have nothing in your house that you do not know to be useful, or believe to be beautiful.'

> **Make the Link**
> You can take inspiration from famous designers or design movements when designing.

During the Bauhaus movement, designs became more simple and a unique style emerged. Ludwig Mies van der Rohe was an architect, designer and teacher and spoke of his designs, such as his Barcelona Chair, as having a 'less is more' approach.

Design movements are periods of time that have a particular style, philosophy, functionality or other developments. The key design movements from the past century are: Arts and Crafts; Art Nouveau; Art Deco; Bauhaus; Modernism; Memphis; and Post Modernism.

AN INTRODUCTION TO DESIGNING

❝Less is more.❞

Ludwig Mies van der Rohe (German-born architect, designer and teacher with great ideas for society, 1886–1969)

❝I think it's really important to design things with a kind of personality.❞

Marc Newson (inspirational modern product designer of today, born 1963)

❝I would think twice about designing stuff for which there was no need and which didn't endure.❞

Robin Day (British designer famous for inventing mass manufactured chairs, 1915–2010)

DESIGN

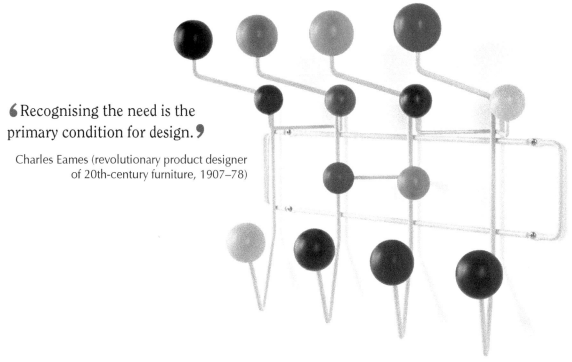

Recognising the need is the primary condition for design.

Charles Eames (revolutionary product designer of 20th-century furniture, 1907–78)

Design activities

This diagram shows an overview of design activities and their relationships with one another.

Make the Link

Chapter 3 explains these activities in more detail.

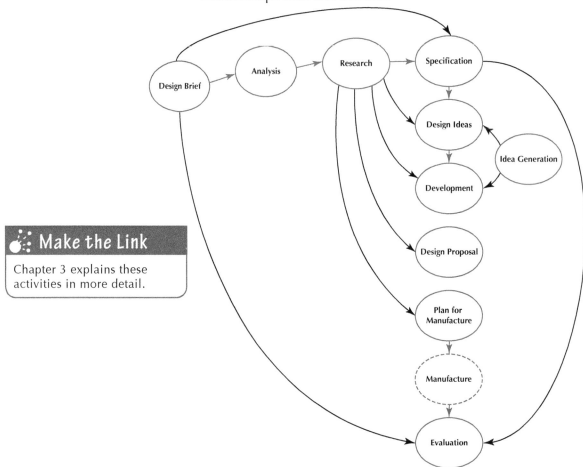

Presenting and recording work

A folder or **portfolio of work** is typical evidence of these design activities; however, each activity is recorded in various ways, from written pieces of work, to drawings and sketches and 3D models. You may, for example, wish to complete an A4 research report, whilst your design development sketches/model photographs/notes may be recorded on A3 paper. This will provide evidence of your thought processes, your decisions and your creativity whilst allowing you to explain and justify the reasons why the final design is the best possible for the brief.

It is very important to record evidence of every design decision using an appropriate method. This may involve using a digital camera to record the work that can't easily be presented on paper.

> **Hint**
> Keep your folio organised by making a checklist for the front.

> **Make the Link**
> English – Keeping a portfolio of written and verbal work.
>
> Graphic Communication – Keeping a variety of drawings and sketches clean and tidy in a folder.

The design team

For every design project there is a team of specialists, all of whom contribute towards the final product.

DESIGN

Key members of a design team

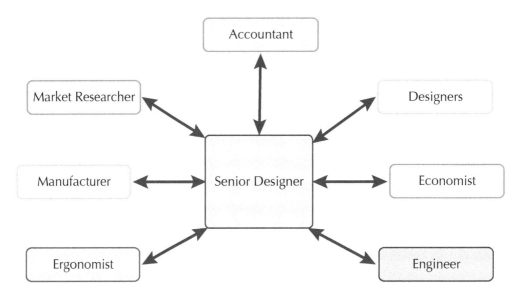

A design team can be a very busy work environment; however, some members can at times work from home, be out doing field research or visiting manufacturers. Each design company will have its own working practices, hours of work and office layout. Some even sit on beanbags.

Make the Link

You will find out more about engineers and manufacturers in the Materials and Manufacturing Unit.

- **Accountant** – organises finances, controls cash-flow, organises staff wages to be paid and generally takes care of the project's incomings and outgoings.
- The **designers** – develop the product.
- **Market researcher** – investigates current products on the market, market trends, the brand identity and potential success of the product.
- **Economist** – determines the correct selling price for the product.
- **Engineer** – works with the technical elements of the project. They may be experts in electronics or material limitations or manufacturing techniques, for example.
- **Ergonomist** – researches human limitations to make the design suitable for use.
- **Manufacturer** – makes the product. They are experts in producing and assembling the parts of the product in the most efficient way.
- **Senior designer** – manages the design team and communicates with the client.

Large companies can employ thousands of people in just one design team. The team may work on one product, but could potentially be developing several products at once.

The design team for a car, for example, has sub-teams with their own designing or manufacturing expertise, for example in mechanics, interior fittings or electronics. Within these sub-teams will be specialists like ergonomists, designers and engineers.

Some design studios, however, are small, with design team members having more than one role within a company, or they may contract other agencies to help them such as a market research company to conduct their research, or employ a manufacturer to advise them on the most suitable manufacturing methods.

Design team communication

Communication is very important when working in a design team. Regular contact between design-team members allows the team to monitor the progress made, resolve any issues with the design, share their ideas, continually evaluate the design, contribute specialist knowledge and also discuss how the design can be improved. Also, the key members of the design team must meet with the client to discuss the progress of the design and ensure that the team is on track to create what the client is looking for.

DESIGN

Other roles that contribute towards designing products

Subcontractors can often be used to carry out specialist tasks, as their talent is required to make the project a success. They may be graphic designers, psychologists or specialists in their field, such as Early Years Workers helping to design toys.

Additionally, there are other people who contribute towards designing products:

The retailer, who sells the product online or in a store, can provide input to the design team, such as sales information or by helping to generate demand for a product through store promotion. The consumer, who uses the product, has some links with the design team in that they create the demand for the product, and can provide feedback and useful marketing information through warranties.

Other members of the design team may include materials specialists, packaging specialists, model makers, product testers and environmental specialists.

> **Make the Link**
>
> Consumer demand, (see page 33), is created through careful market research.

Activity

The design team placemat

This is a group task with groups of up to four members. The activity is to be completed on A3 or flipchart paper.

A design idea for a child's balance bike is shown below:

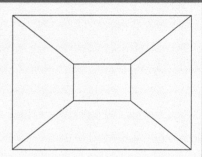

1. Draw the placemat: Divide the paper into segments based on the number of members in the group, leaving a box in the middle. The example shown is for a group of four.
2. As a team, choose **one** design team member and write their job title in the middle box of the placemat.
3. **2-minute timed task:** On your own and in your own segment of the placemat, write the tasks which that design team member will do when working on the design of the child's balance bike.
4. **4-minute timed task:** Now, as a team, discuss and agree which four main skills and tasks the design team member must complete. Write these in the centre box.
5. **4-minute timed task:** As a group, prepare to feed back your answers.

AN INTRODUCTION TO DESIGNING

🔍 Design team case study

Real Wood Studios

Graham Murdoch obtained a degree in Product Design at the Robert Gordon University in Aberdeen. He then attended the City of Glasgow College to study Furniture Making, while also working part-time in a furniture-making workshop. He set up business at Real Wood Studios, a cooperatively-owned workshop based in the Scottish Borders, in September 2004. He produces a vast range of wooden products; from jewellery boxes to reception desks.

Graham is responsible for designing and manufacturing his furniture, while also meeting with his clients, and acting as the economist and ergonomist for each project. He also conducts his own market research, along with many other tasks that arise each day.

'Commissioning bespoke furniture involves many key stages, from discussing the initial brief with a client and generating the first draft of a design, to selecting the right timber for each component. These stages are just as important as the construction of the piece itself.'

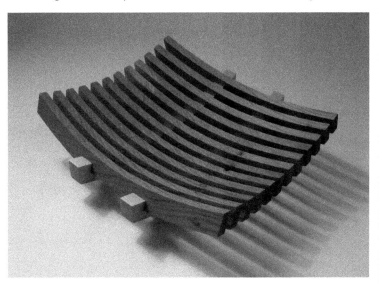

Graham's designs consist of simple forms using clean lines and sweeping curves and often combine two contrasting timbers.

✔ Test your knowledge

A design team have worked together to create a snowboard.

Link to suggested answers
www.collins.co.uk/pages/scottish-curriculum-free-resources

1. State the member of the design team responsible for researching human sizes to make the snowboard fit the target market.

2. Explain why the manufacturer has to work closely with the engineer.

3. Describe why it is important that the design team keeps in touch and updates the client regularly.

4. A graphic designer was used to create a top-sheet design for the snowboard. Explain one benefit of sub-contracting this aspect of the design.

DESIGN

> **Make the Link**
>
> The products we use every day satisfy either our *wants* or our *needs* (see pages 17–18).

Commercial products

Commercial products are those products that are sold to make a profit for an organisation. They are items that we use in all aspects of daily life, such as trainers, watches, coffee machines, bike locks, bins, fruit bowls, chairs, folders, showers, shelves, radios and so on.

Activity

Commercial products

This is a whole class task. The activity can be completed on a whiteboard or maybe through class discussion.

1. On your own, think of your daily routine. In your head, take yourself through your day so far. Visualise the rooms that you were in this morning: bedroom … bathroom … kitchen … Consider your journey to school. Did you stop anywhere on the way?
2. Now consider the commercial products you used along this journey. Share your answers with the class.

> **Hint**
>
> Use the answers from this activity to provide a range of interesting products for future class discussions.

Brand names

Some commercial products are so successful that the brand name has taken over the functional name of the product. For example, a vacuum cleaner is often referred to as a Hoover, even if the vacuum cleaner is not manufactured by Hoover. Our use of brand names extends even further into our language as we adapt them to suit our vocabulary, such as 'doing the hoovering'. Other brand names we have adopted into our language are Tannoy, Aspirin, Astro Turf, Hi-lighter, Post-it, Escalator and Sellotape.

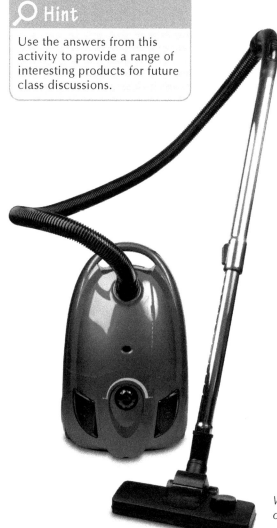

Would you refer to this as a Hoover or a vacuum cleaner?

AN INTRODUCTION TO DESIGNING

Needs and wants

There are products we **need** and products we **want**.

Products we *need* are those that fulfil our basic needs as human beings. There are products we need to survive, such as food and water, and products we need to keep us safe, such as smoke alarms.

Needs

Products that satisfy our needs include products that:

- give us shelter
- provide warmth
- protect us from harm
- encourage hygiene
- make us feel valued and appreciated.

A tent is an ideal way of providing shelter, warmth and protection. Tents are a main residence for many people across the world and are often provided as shelter after natural disasters.

Wants

Products we *want* are those that fulfil our desires as human beings. Products we want to enrich our lives include TVs and roller blades; products we want to make our lives easier include food blenders and smartphones. Products that satisfy our wants include products that:

- make us feel socially accepted
- enhance our comfort
- keep us up to date or informed
- make us fashionable
- improve our health and wellbeing
- make us happy.

*We may **want** to own a TV, but we don't **need** to.*

Check your progress

I can:

	HELP NEEDED	GETTING THERE	CONFIDENT
describe the overview of design activities	◯	◯	◯
describe the members of a design team and their roles	◯	◯	◯
explain the term 'commercial products'	◯	◯	◯
explain the difference between needs and wants.	◯	◯	◯

2 The factors that influence design

By the end of this chapter you should be able to:
- describe the functional factors that influence design of products
- describe the performance factors that influence design of products
- describe the marketing factors that influence design of products
- describe the aesthetic factors that influence design of products
- describe the ergonomic factors that influence design of products
- explain the ways in which sustainability influences design decisions.

The factors that influence design

There are many different factors that a designer must consider to ensure the final product is a success. The designer has to balance these factors carefully, so that the design proposal meets the design brief and specification.

Prioritising factors

The factors that influence design are:

- **Function**
- **Performance**
- **Market**
- **Aesthetics**
- **Ergonomics**

There is no order of importance with design factors, as each individual product has its own design priorities. For example, the most important factor when designing a tin opener is *function* as it must open tins, however the designer must also consider the *performance* of the tin opener as it must be reliable, and *ergonomics* as the tin opener must be comfortable and safe to use.

Design and sustainability

Designers also have a responsibility to ensure that their products are **sustainable**. This is not a single design factor, but an issue that impacts and influences design decisions regarding all five factors.

Make the Link

Sustainability is covered on page 51.

Function

Function is the purpose of a product; or, to put it simply, function is the job the product is designed to do. A lawnmower cuts grass, a washing machine washes clothes and a pen writes on paper. These are all functions.

When products do not fulfil their function, they can be very frustrating for the user and are a waste of money for the manufacturer, retailer and the consumer.

Primary function

The **primary function** is the most important function of the product. If a product does not fulfil the purpose for which it is intended, its primary function, then it is useless. Imagine a shoe rack that is not big enough to fit any shoes on it, not even enough space for a flip flop! The shoe rack would not fulfil its primary function, which is to hold shoes.

Secondary function

Many products have additional functions. These **secondary functions** are extra features or ways in which the product can be used, making the product more attractive to consumers. Imagine the shoe rack again. This time it cannot only hold shoes, but it also has hooks on it for hanging jewellery or belts. It has one primary function, which is to store shoes, with the secondary function of storage to appeal to consumers.

THE FACTORS THAT INFLUENCE DESIGN

✔ Test your knowledge

Primary and secondary functions

Link to suggested answers
www.collins.co.uk/pages/scottish-curriculum-free-resources

1. Explain the meaning of the term 'primary function'.
2. Explain the meaning of the term 'secondary function'.

A dining table is shown on the right.

3. State the primary function of the dining table.
4. Describe a secondary function of the dining table.
5. State five products with secondary functions.

Performance

The **performance** of a product is how well a product works. For example, you would expect a disposable razor to shave, but not to a high quality for a long period of time. Neither would it be too surprising if it cut you, whereas an electric razor would shave to a higher quality, would be easy to use, comfortable against your skin and you would be very surprised if it cut you. Both products perform the same function but to a different degree of success *or* level of performance.

Performance and conditions of use

The performance of a product can depend on external conditions. Lack of power, the level of light, the wrong temperature and the frequency of use are just some of these conditions. For example, a solar-powered garden light gains power from the sun and so, in this case, the level of light it receives influences its performance.

In extremely cold temperatures, some electrical products do not work at all. LCD (Liquid Crystal Display) screens on smartphones struggle to function below −14°C as the reaction time of the sensors decreases as the temperature decreases. This poor performance is due to a very simple reason: the LCD screen is a liquid and it freezes.

This solar-powered lamp has a solar panel as a secondary power source that extends the battery life by 50%.

> **Hint**
> Solve bad design by thinking about products that are not easy to use. Consider the frustrations you have with them and then think about what would make them easier to use.

> **Make the Link**
> Groups of people who use a specific product are called a target market.

Performance and ease of use

The ease of use of a product is how *simple* it is to operate. Products should not be overly complex or difficult to understand. They should not be awkwardly shaped or exhausting to use. When a product is difficult to use, we can feel frustrated, confused, angry and even defeated.

It is the designer's job to understand the user and their limitations, such as their age, gender, physical ability and if they have any disabilities. Every person is unique and will react differently to products.

A common product that can be difficult to use is scissors. There are many different types of scissors that are designed for different groups of people or different *markets*. For example, children have their own small blunted scissors, left-handed people have left-handed scissors with the blade reversed so they can see the cutting line, and there are specialist spring-loaded scissors that ease the workload on the fingers of arthritis sufferers.

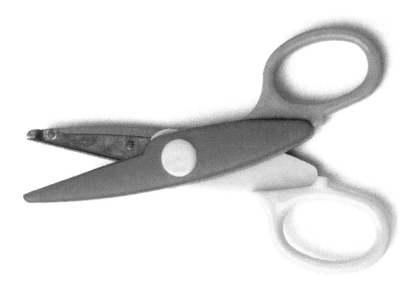

Performance and fitness for purpose

The term *fitness for purpose* is often used to describe how well a product performs.

A salt grinder is the correct size for a hand, is able to store a reasonable volume of salt, can easily be refilled, is easy to clean as it can be taken apart and it also has a grinding mechanism that is made from a material that will not be corroded by the salt. The salt grinder is therefore demonstrating *fitness for purpose*.

Performance and ease of maintenance

The **maintenance** of a product is how the product is repaired once it is broken or **maintained** to keep it in safe and good working order. Maintenance of products should be easy, otherwise the product may be thrown away, as consumers are easily persuaded to replace products with new ones.

Some products last for years before they require any maintenance and others may only last a few uses. Products need to be repaired for many reasons:

- Parts wear away and need to be replaced.
- Materials become damaged and need to be replaced or repaired.
- Joints, screws or fixings become loose and need to be tightened, replaced or re-glued.
- Electrical components fail and need to be replaced.
- Paint, varnish or other finishes wear away and need to be reapplied.
- Mechanical parts need to be serviced, oiled or cleaned to operate.

If the quality of construction/manufacture is not good enough, a product can fall apart quickly. The need for products to be repaired can also mean the need for a specialist spare part or repair person, which may only be available directly from the manufacturer. This adds to the overall price of the product.

Some ski and snowboard goggles have an interchangeable lens design that lets you change the lens for different snow conditions or in the event of damage.

Make the Link

Design for disassembly is a key part of the cradle-to-cradle concept, which promotes durability and availability of replacement parts (see page 52).

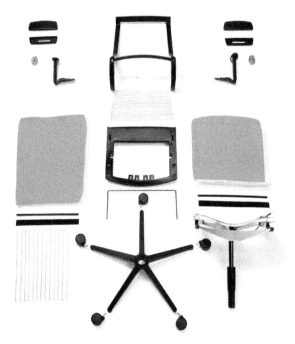

DESIGN

Cradle to Cradle

Products that cannot be repaired create waste when they or their components are thrown away and are not reused or recycled.

Designers, therefore, have a responsibility to design products that can be repaired easily or that can be taken apart for ease of reusing or recycling the component parts. The Think Chair by Steelcase (left) was one of the first products to qualify for **Cradle-to-Cradle certification** as it can be disassembled in 5 minutes with common tools, allowing for parts to be easily replaced if necessary.

Performance and durability

The **durability** of a product is its ability to withstand wear and tear. Consumers trust products to be durable. Ideally, we would like all the products we buy to be durable, to work well every time and to last for a long period of time without becoming weak and breaking. Realistically, this is very difficult as it requires them to be:

> **Hint**
> Durability means that a product has a **dur**able **ability**.

- waterproof
- shatterproof
- heat resistant
- shock absorbant
- sharp
- hard wearing
- strong
- long lasting
- tough
- sturdy
- robust.

The rubber stoppers on the end of the metal legs provide extra durability for the chair.

Creating a durable product involves:

- Selecting hard-wearing materials and using resilient manufacturing processes.
- Using robust assembly techniques.
- Understanding how the product is used.
- Considering the conditions of use (hot, cold, indoors, outdoors, etc.).
- Knowing how the target market (the user) will treat the product.

> **Make the Link**
> Varnish, wax, oil, paint and plastic dip-coating are material finishes that can prolong the life of a material.

THE FACTORS THAT INFLUENCE DESIGN

Metal and wood may require a finish or treatment to make them durable in certain conditions. For example, a pine bird table can be varnished, which will protect the pine from being damaged by the weather. Plastic does not require a finish as it is waterproof and is, therefore, resistant to most weather conditions.

Make the Link
Durability is a property of many materials.

Performance and lifespan

The lifespan of the material will have an impact on the lifespan of the product. Some products are designed to be used just once, such as a plastic bottle of water.

Filling up the bottle and reusing it is great for the environment but eventually, through constant use, the thin plastic walls will start to weaken.

A more durable and sustainable option is a metal water bottle, resisting wear and being able to withstand greater impact than a plastic bottle. The metal water bottle is made from aluminium; extremely lightweight but very strong.

Materials can become damaged with constant use or misuse. Sometimes, they are damaged by weather conditions, such as a wooden shed rotting away.

Hint
Try to use lots of different materials throughout the course to develop your knowledge of different materials.

Make the Link
Disposable plastic products are bad for the environment.

The way that a product is built will impact on how well it performs. If it is built using manufacture methods that make it strong and durable, then it will have a longer lifespan and will be able to endure more rigorous use.

When a product is made from more than one component, it must be assembled in a way that allows for the components to join neatly and accurately. Consumers will notice if the product is not assembled well and this will have a negative impact on their confidence in the product when using it. A poorly built product will probably not function efficiently and may even cause injury to the user if components break off or if fittings protrude.

Make the Link
Commercial manufacturing is examined in Chapter 6.

DESIGN

Designers no longer create large, boxy electronics like this 1970s TV. Developments in electronics now allow smaller component parts to be used.

Size

The size of a product impacts on the performance, whether large or small. Smaller microchips and electronic components have allowed technological products to shrink over the past few decades. Products that get smaller and slimmer are said to miniaturise. This is to make products lighter and generally more comfortable to use.

Making products smaller, however, can have a negative impact on their performance. For example, a small, lightweight chain lock for a bike might seem like a good solution to the problem of carrying a big, heavy lock around. However, the smaller version may be easily cut off and so offers less security.

Similarly, some electronic products have reduced so much in size that the buttons have become too small for fingers to operate them. The size of a successful product, therefore, requires consideration of ergonomics and function.

Make the Link

Deciding on a product's size requires ergonomic data. Designers must consider human limitations when designing.

THE FACTORS THAT INFLUENCE DESIGN

Safety issues associated with performance

The safety issues associated with the performance of products include:

- poor construction leading to products breaking and users being injured by broken products
- durability of materials used not meeting the needs of the user and parts wearing away, reducing the safety of the product
- component parts not being shatterproof
- electrical parts failing and creating safety hazards through the product not working or the electrical components themselves causing injuries such as electric shocks
- mechanical parts breaking, which can cause accidents through improper use of the product.

✔ Test your knowledge

Performance

Battery-powered milk frothers have become a popular way to have a frothy coffee in the comfort of your own home.

> Link to suggested answers
> www.collins.co.uk/pages/scottish-curriculum-free-resources

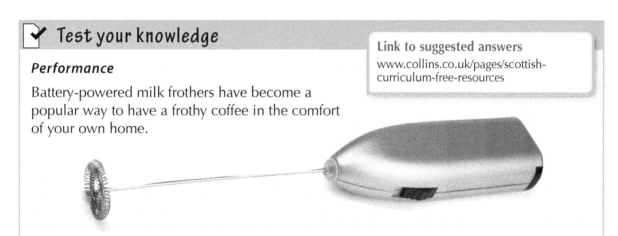

1. Explain the meaning of the term 'ease of use'.
2. With reference to the milk frother, explain its ease of use.
3. Explain the meaning of the term 'durability'.
4. The conditions of use will affect the durability of the milk frother. Explain one method of maintaining the milk frother.
5. Explain **two** additional durability issues relating to the milk frother.
6. Explain the meaning of the term 'ease of maintenance'.
7. Describe **one** reason why the milk frother is easy to maintain in terms of:
   a) materials
   b) joining methods
   c) replacement parts.

Market

The market describes the consumers who purchase the product. Understanding the market is important as it will determine the success of the product in terms of its popularity. Marketing involves carrying out research with the people who will typically use the product, promoting and selling products.

Target market

The **target market** is the group of people who use the product. The target market can be defined by age, gender, interests, lifestyle and location, along with many, many other aspects.

The target market for a product may be broad and general; a pair of mid-priced women's gym trainers is aimed at women who exercise but who don't spend a lot on their gym wear.

The target market for a product may be very specific and select; a pair of high-priced, super lightweight, women's spiked running shoes is aimed at a professional athlete who depends on the product to perform.

A market can be split by cost into low, mid, high and luxury. Consumers on a tight budget, for example, would be more likely to purchase low-cost products.

A good designer will carry out market research to find out what the target market wants or what issues exist. Market researchers can use techniques such as focus groups, consumer reviews and user trips to find out what the target market want from their product. The designer can then take this information 'back to the drawing board' to design a product that the target market will want.

A focus group discusses the launch of a new product.

Introduction of new products

New products are introduced to the market when new technology emerges. When older technology is replaced by new technology, which is faster, smaller and easier to use, this is called **product evolution**. Usually, the function of the product is updated or improved, and the aesthetic may be modernised. An example of this is seen in the way we watch movies: DVD players replaced VHS video players in the late 1990s; but before VHS players, there were Beetamax players and Cinefilm predated those. These changes are made possible by technological advances, new materials and developments in manufacturing processes. Alternatively, products can be introduced that are entirely new to the market and do not replace a previous product, such as satellite navigation products.

Marketing techniques to influence sales

When a product is introduced to market there are a number of techniques that can be used to help grow a customer base. These include:

- Buy one get one free (BOGOF)/bulk buy offers.
- Low introductory price/discount deals on specialist websites.
- Adverts, e.g. TV, radio, billboard etc.
- Promotional prize draws/free gifts.
- Celebrity endorsement.
- Social media advertising/presence.
- Product placement on TV.
- Selling products under a big-brand name.

DESIGN

Social expectations

Today, people in the developed world consume products at a worrying rate. Products are advertised as life-changing; the solution to many small issues in everyday life, but good quality design can improve the quality of all our lives.

The main **social expectations** from a new product are:

- It has new or better features.
- The style and look are improved.
- It is better value for money.
- There are more options or choices.
- It is more sustainable or environmentally friendly.

Here a focus group gives feedback on tropical fruit-scented shampoo. The designer is asking the group some questions to investigate creative ideas for bottle shapes.

THE FACTORS THAT INFLUENCE DESIGN

Social expectations continually change, and new technology and new inventions emerge constantly. For example, we saw the laptop shrink in size to become a netbook. Then it lost its keyboard and became a tablet. We expect it will shrink a little more, improve its features, improve its quality and change its style.

✔ Test your knowledge

Link to suggested answers
www.collins.co.uk/pages/scottish-curriculum-free-resources

Marketing

The Julian chair, designed by Javier Mariscal, is:

- made from plastic
- available in red, green, yellow and white finishes
- suitable for outdoor use
- £80·00.

1. Explain the meaning of the term 'target market'.
2. Describe the main target market for this product.
3. With reference to the Julian chair, explain the meaning of the term 'market segment'.

The marketing mix

There is normally a **marketing strategy** for every commercial product. This is a plan that is written to define the marketing aim for the product; where it will be sold, who it is intended for and why it should make a profit. It is useful for the designer to know about this strategy as it will give them a deeper understanding of the product.

The **Four Ps** of marketing are product, price, place and promotion. This is called the **marketing mix**. Making the right choices in each of these categories is the basis for success.

- **Product** – the item for sale. The product must satisfy a *want* or *need* if it is to be successful. The product will be defined by its features and extra benefits to give it a unique selling point (USP) and to make it interesting to consumers. It must have a place within current product ranges or fit into a gap in the market, selling among other brands.

- **Price** – the selling price of the product. The impression the price gives is important. While it must make a profit, the price can make a statement about the product. Expensive products can appeal to consumers as they are regarded as luxury items. The price must set a tone and make the right impression to attract the target market.

- **Place** – where the product is available to the consumer. The buying experience itself is important, from the store location to the 'welcome' from the staff. When online shopping, ease of use and ability to seek assistance are vital to the buying experience. The place makes a statement about the consumer's social status. Place also means where the product is advertised to the consumer, perhaps on specific TV channels, via the internet, pop-up shops, or at specific times of the day on the radio.

- **Promotion** – advertising and informing consumers. Keeping consumers informed and up to date with new products makes them aware of their choices. Clever advertising will push products onto unsuspecting consumers. Other consumers are harder to win over as they rely on details and demonstrations before they make a trusted purchase.

> **Hint**
> An effective marketing strategy will combine all of the Four Ps in the marketing mix.

Branding

A **brand** is a product or company identity. It is how we recognise commercial products instantly with our senses, through colours, textures, sounds, smells or tastes. A brand is not a logo; it is a psychological link between you and the product. Brands play an important role in marketing as they make a link between the product and the person.

People actually build relationships with brands. They *love* brands. They are *protective* of their favoured brands. They *trust* brands.

In some products, the brand can be quite discreet. However, in others you will notice the brand instantly. The matt black texture and gold ceramic plates on GHD hair straighteners or the curvy and compact shape of the Volkswagen Beetle make the brand instantly recognisable without the need for any logo.

Actually, there really is no need for a logo on a product. However, it is usually added to keep the company or product name in your head, reminding you of the brand and enhancing your loyalty to it. Placing a logo on a product is also a way to advertise the brand.

> **Hint**
>
> Think about the last time you visited a fast-food outlet. Think about the overall quality of the experience. What words come to mind? Compare that to the last time you visited a restaurant. What words would you use to describe this experience? These words help you to form a mental image of the company. This is the brand.

Consumer demand

'**Consumer demand**' is a term used to describe the popularity of a product. Products with a high consumer demand are sought after; perhaps due to fashion, new technology, new design or basic necessity. Products such as toothpaste, water bottles and shoes have high consumer demand as they wear away or run out and are needed by millions of people on a daily basis.

Consumers often feel loyal to particular brands as they have bought reliable and quality products from them previously. This is a way of ensuring that consumer demand for products can be maintained. Products that are in high demand are mass manufactured to ensure that demand can be met. Common products, such as plastic water bottles, can be produced continually, 24 hours a day, seven days a week, to ensure that consumer demand is met.

> **Make the Link**
>
> The types of commercial manufacturing methods explained in Chapter 6 are employed depending on the different levels of consumer demand for the product.

Amazon has maintained consumer demand through successful branding, teamed with good value and good quality, while the newest Kindle models have improvements that appeal to its network of consumers.

Methods to support sustainability

Technology push

'**Technology push**' is a term used to describe how a product comes to market, based on new technologies, new materials and new manufacturing methods, rather than in response to consumer demand. When a new product becomes available to buy, a strong advertising campaign is needed to create consumer demand. To launch a new product is a great risk, as the client will lose money if the product doesn't sell.

In the past 50 years the world has evolved into a fast-paced, highly technical and technological place. We constantly experience new technology that is smaller, faster, smoother, quieter, stronger and generally more capable. Technology push may involve an improvement on current technology or simply something totally new.

> When the Nintendo Wii was launched in 2006, the advertising campaign was so strong that the demand was extremely high and it was almost impossible to find anywhere to buy one. Shops had month-long waiting lists of hundreds of people and it was equally impossible to buy the product online. Nintendo was very successful in creating a buzz around what was, at the time, amazing new wireless-control technology for a games console. However, their manufacturers couldn't keep up with demand.

Smartphones are constantly updated with new technology, and consumers can benefit from new features and functions each time they upgrade their handset.

The large range of bagless vacuums on offer these days is a result of market pull.

THE FACTORS THAT INFLUENCE DESIGN

> ### Activity
>
> *Marketing and branding teamwork*
>
> This is a group task with groups of up to four members. The activity is to be completed in a jotter/sketchbook, on paper or can be recorded electronically.
>
> The time for each task is **four minutes**. Form small groups with up to four members.
>
> 1. List as many *brand names* as you can.
> 2. Discuss the meaning of the term 'product branding' and then write the definition.
> 3. List as many examples of products with a high consumer demand as you can.
> 4. Discuss the meaning of the term 'consumer demand' and then write the definition.
> 5. List as many products with *market pull* as you can.
> 6. Discuss the meaning of the term 'technology push' and then write the definition.

Market pull

Market pull means there is a demand for a *product* and so a product is designed to fill that need. When bagless vacuum cleaners went on sale they were very expensive and not all households could afford one. A gap in the market opened up for more affordable bagless vacuum cleaners and so the market became wider and more affordable. This is market pull.

> **Hint**
>
> 'Market pull' means the market is pulling the designers and manufacturers towards designing a product for them.

Aesthetics

Aesthetics is the way a product is perceived by our senses; its look, taste, smell, tactile quality (feel) and sound.

Typically, aesthetics in design and manufacture refers to the shape, proportion, size, colour, contrast, harmony, texture, materials, fashion of products.

> **Hint**
>
> Words to avoid using are **cool**, **nice**, **good**, **bad** and **OK**. These words are simply not creative enough. A professional designer would not give a presentation about a 'nice' product.

Shape

The shape of a product must be attractive to the eye, while also being functional and ergonomical. For example, a hairdryer has a stereotypical shape, which is composed of a comfortably shaped handle, a cylindrical housing for the motor/fan and a tapered section for the heating element.

The shape of each element of the hairdryer will have unique design considerations that the designer has to resolve. For instance, when developing the shape of the handle, the designer must apply anthropometric data to ensure that the product will comfortably fit the hand and yet be shaped to suit the target market needs, whether it be a sleek, slender shape or a curvy, flowing shape.

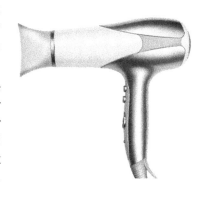

Colour

Colour is usually the first detail that the eye notices. To appeal to a wider market, some products are available in a variety of colours and product colours can be updated frequently in line with fashion trends.

Colour can help to bring a product to life, help us understand how it works (such as in colour coding; a green button for 'on') or even evoke an emotional response.

Colour theory is important in product design. We automatically respond to colours on a subliminal level, learning the meaning of colours without even realising it. For example, we interpret a red flashing light on a product as a warning light or low-battery signal, without even looking at a manual or the instructions. This link between colours is a result of product psychology.

To encourage an emotional connection to the product, these aspects of colour theory are used in design:

- **Black** is the colour of authority and power (also of death in most cultures). It is stylish and timeless, always in fashion.

- **White** is the colour of innocence and purity. It reflects light and is also considered to be timeless. It works well with every other colour. It is said to be neutral, blank and plain. White is often used in design, however it shows dirt and is difficult to keep clean.

- **Red** is the colour of love, heat and passion. It is often used in products that are bold, to make a statement. However, it is also the colour of risk and danger, and signifies 'stop', which can bring negative associations.

- **Pink** is a romantic color, which creates a sense of calm and serenity. It is feminine and youthful, with a sense of softness and delicacy.

- **Blue** represents open space, sea, air and sky, bringing peace and tranquility. But blue can also be sad and cold. Darker tones of blue are stronger; they represent trust, and signify loyalty and security. Navy blue represents rules and regulations.

- **Green** symbolises nature, safety and hygiene. It is the colour that is easiest on the eye. Hospitals often use green because it is a calming colour. Green is often positive; it is the colour of 'go' and 'on', but can also represent jealousy.

Red is a colour associated with love but it can also be associated with danger.

- **Yellow** is an attention grabber that lifts moods, encourages focus and promotes positivity. Used correctly, it is uplifting and happy; however, bright yellow products can cause tempers to flare and babies to cry! Writing pads come in yellow as the colour can increase creative thought, but is the most difficult colour for the eye.

- **Purple** is the colour of royalty, luxury, wealth and sophistication. It is soothing and calming, which boosts imagination and creativity. It can be both hot and cold.

- **Brown** is the color of earth and is abundant in nature. It is a natural colour, which represents simplicity and dependability. Brown products seem organic and wholesome, creating stability and grounding.

- **Orange** is the colour of energy, warmth and flamboyance. Its association with citrus fruit makes links with health and wellbeing. It has strength, but is also fun.

This coat rack uses colour to brighten the product. The flip-out hooks are painted in a rainbow of colours to bring a childish tone to the aesthetic, while the white panel helps to balance the variety of colours.

DESIGN

Material

All materials have unique characteristics of colour, texture, pattern, form and quality. Walnut, for example, is an expensive hardwood with a very complex wood grain and is sometimes used in car dashboards to give a luxurious finish.

This stainless-steel worktop has a hard and sterile appearance, but is shiny and hygienic. This makes it functional in a professional kitchen.

> **Make the Link**
>
> Select a material that is both fit for purpose and aesthetically suited to the design by researching material properties in Chapter 6.

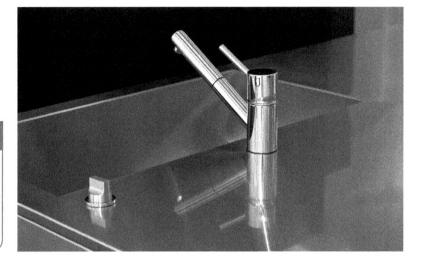

Varnishing is a functional way of showing the natural grain of wood and protecting the material from wear. Materials can bring unique aesthetic qualities such as the grain of a marble table, the surface pattern of zinc-plated garden furniture or the transparency of an acrylic picture frame.

This stainless-steel worktop has a hard and sterile material aesthetic that makes it look shiny and hygienic.

THE FACTORS THAT INFLUENCE DESIGN

Texture

The **texture** of a product describes the surface qualities of a material, which we experience through touch and sight. We may be drawn towards products with particular textures.

Sections of a product may have contrasting textures – to highlight operating controls, for example. In the case of the oven shown below, the ridged texture of the controls contrasts against the smooth shiny metal, making the buttons stand out.

Many modern manufacturing methods can achieve such a high level of intricate detail that it is possible for a texture to be imprinted onto products or components during the manufacturing process. Products may have raised patterns, shapes or a company's logo imprinted on them.

> **Make the Link**
>
> Texture is not only used to enhance the aesthetic appeal of products. It can ergonomically benefit the user in bringing comfort, improve the performance of a product (e.g. with grip) or assist the ease of use.

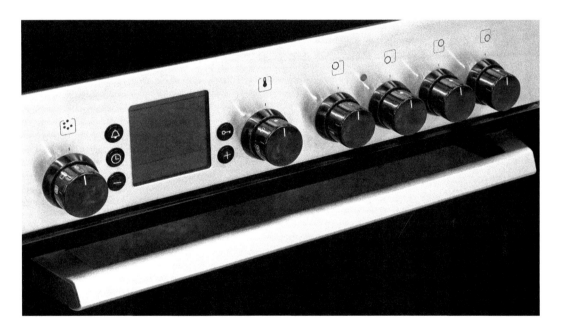

Proportion

The proportion of a product is the scale and size of the product, or the relative sizes and placement of the product parts. When all parts are in harmony with each other, they create unity in the aesthetic of the product.

In this asymmetric wall shelf, the proportion is well balanced: both of the shelves are the same length, and the width and thickness of the wood is the same throughout. The holes for the wall fixings are offset,

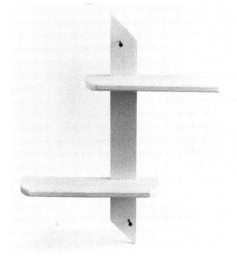

but – as they are the same distance from the edges of the material – they are well placed proportionally. This wall shelf shows that design does not need to be symmetrical to be proportional.

Test your knowledge

Link to suggested answers
www.collins.co.uk/pages/scottish-curriculum-free-resources

Aesthetics

The pencil sharpener desk tidy shown on the right is designed to hold pens and pencils, while keeping your desk looking organised.

1. Describe the aesthetic qualities of the pencil sharpener desk tidy in terms of:
   a) shape
   b) proportion.
2. Describe **two** ways in which colour could improve the aesthetic of the pencil sharpener desk tidy.
3. Describe **two** ways in which the materials enhance the aesthetic of the pencil sharpener desk tidy.

Harmony and contrast

A harmonious product is one that is well balanced and gives a calm feeling. It has soothing textures, proportions, materials or colours and a neutral aesthetic. Products made from one material and in one all-over colour can also create this sense of calm and comfort.

A product that has contrast is one that involves opposite textures, materials or colours. It is designed to draw your attention to a particular part of the product or to make the design stand out.

THE FACTORS THAT INFLUENCE DESIGN

The colour wheel can be used to select harmonising colours. These are the colours that are directly beside each other. It can also be used to select contrasting colours. These are the colours that are directly opposite each other.

This chair uses contrasting shapes and materials, which give it a unique style.

This sofa, in shades of green, has a sense of harmony in its colour scheme.

Fashion

Fashion changes over time and describes, at a particular moment, the current styles, colours, shapes, textures and technologies in our clothes, products, homes, and in many other areas of our lives. It can signify culture, gender, age and values. Fashion is a way of stating a social identity and being fashionable meets the basic human need to belong.

It is easy to associate fashion with clothes. However, the products we use go through different fashion trends too. The world of furniture design, for example, has its own fashion shows, called trade shows.

> **Hint**
>
> To find out the current aesthetic styles, look at home-decoration magazines, store catalogues, social media and in stores.

Ergonomics

Ergonomics is about the way that humans interact with products. An ergonomic product will be comfortable to use and will not cause you any injuries or strain when using it. Also, it will not be too loud or make you feel upset.

An ergonomic product is:

- comfortable during use
- easy to use
- simple to understand
- easy to lift, twist, push, pull, etc.
- easy to grip or hold onto
- straightforward to maintain or repair
- safe to use.

The most obvious ergonomically designed products are adjustable chairs, any product with buttons or controls and anything with a handle shaped to fit fingers. However, most products have been designed with some ergonomic consideration.

A basic pen has many obvious ergonomic design factors. It is long enough to fit your hand width, the diameter of the pen makes it easy to hold, there may be a texture or shape that makes it easy to grip, and the lid probably has a different texture or design feature that makes it easy to feel the difference between the two parts and remove the lid without too much concentration.

Ergonomic designs can be complicated, with adjustment levers or features designed to be adjustable to suit the needs of a range of different-sized and shaped people. The designer is tasked to create a design that is both ergonomically and aesthetically appealing to the target market.

The Tip Ton chair is ergonomically designed to tilt forward when you lean in towards your work, then tilt back when you want to sit back.

THE FACTORS THAT INFLUENCE DESIGN

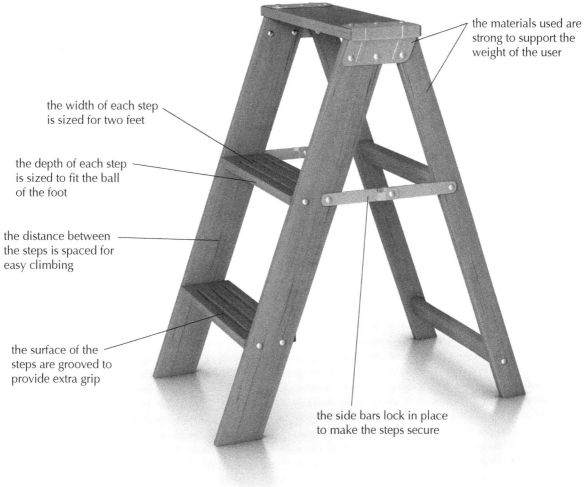

A stepladder has a number of ergonomic considerations that make it accessible for a wide range of people:

- the materials used are strong to support the weight of the user
- the width of each step is sized for two feet
- the depth of each step is sized to fit the ball of the foot
- the distance between the steps is spaced for easy climbing
- the surface of the steps are grooved to provide extra grip
- the side bars lock in place to make the steps secure

Understanding how humans interact with products

Looking at the way in which a product will be used can help you to understand which sizes are important for its design. When designing a kettle, hand size data will be useful to size the handle.

The culture or race of a person can influence anthropometrics. This difference in human sizes means international companies must develop adjustable products or, perhaps, even manufacture an alternative size for a foreign market.

Colour is also important in ergonomics. Products can be colour coded to influence users or communicate something in a split second, perhaps even without the user realising it.

DESIGN

Age and ability can also influence anthropometrics. Big button phones are designed for elderly, disabled and visually impaired people to make calls more easily, as they have extra large, high-visibility buttons that are comfortable to use.

Anthropometrics

To design a product that is comfortable to use, human sizes such as height, arm length, thumb size, etc. need to be researched. The study of human measurements is called **anthropometrics**.

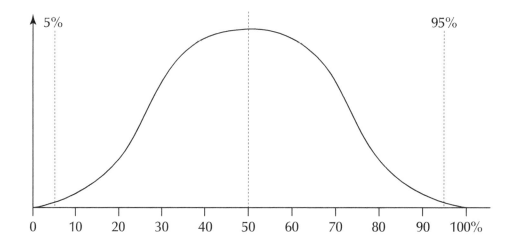

Measurements of human dimensions can be presented in a graph. This graph is called the **bell curve**. In the very centre of the graph is the 50th percentile, which is the most common size. This may be the average hand span, the average hip width or even the average size of a big toe.

The 0 to 5th and 95th to 100th **percentiles** are the less common and most extreme human sizes. Below the 5th percentile are the extremely small human dimensions, while above the 95th percentile are the extremely large human dimensions. Generally, designers do not design products for these extreme human dimensions, as there are very few people in this range. Accounting for these dimensions can make products too big and uncomfortable for the majority of users.

So, designers usually design products that suit 90% of the population. This is cost effective as it means products will meet the needs of the majority of people. The remaining 10% of the population may require to make adjustments and adaptations to the product, or may even require a specialist product. Designing to incorporate the needs of this group adds unnecessary cost for a limited return in sales.

An ergonome is ideal for assessing the dimensions of scale models.

Ergonome

An **ergonome** is a small, scaled-down human model that is used with scale models to test the size of products. If the ergonome is 1:10 then the model dimensions are divided by 10 to make a model that suits the ergonome. Then the ergonome can be placed next to or on the model to test the dimensions of the product.

Establishing critical sizes

Products should be able to be used by 90% of the population. When selecting the best sizes it is wrong to always design for the 50th percentile.

For example, an umbrella designed for the shoulder width of the 50th percentile would mean that 50% of the population would not be sheltered appropriately by the umbrella. The same umbrella designed for the 100th percentile (the whole population) would be much wider, heavier and would use more materials.

To establish the most important sizes for the product you are designing, you should consider:

- the target market
- the function
- safety
- comfort
- ease of use.

> **Make the Link**
>
> Use ergonomes when creating scale models during design development.

> **Hint**
>
> Consider the length, breadth, thickness and weight of all the component parts of your design to make the product a suitable size.

DESIGN

✓ Test your knowledge

Anthropometrics

Games controllers are designed with human sizes in mind.

Link to suggested answers
www.collins.co.uk/pages/scottish-curriculum-free-resources

1. Explain how the following anthropometric data has been applied to make the design of the games controller comfortable to use:

   a) thumb length

   b) hand length.

2. State and justify the percentile range that would be used when selecting anthropometric data for the games controller.

Physiology

> **🔍 Hint**
>
> Human physiology determines the key physical movements that are studied in ergonomics: pushing, pulling, twisting, turning, lifting, grabbing, gripping, holding, pressing, squeezing, etc.

Products must meet the capabilities of the human body and so designers must understand physical ability as we all have limitations when it comes to pushing, pulling, twisting, turning and lifting. This is called **physiology**. Take a sports drinking bottle for example. It should be designed in such a way that any person can easily twist the top to open it. Everyone has a different set of hands but the bottle should be easy for any shape and size of hand to grip and lift it.

Repetitive actions can cause muscle fatigue in your hands and fingers. Ideally, use of products should not cause muscle strain or mental tiredness, which is known to reduce reaction/response times.

Psychology

Psychology is the study of the human mind and behaviour. Designers must understand human behaviour to design appropriate products. For example, we are naturally sensitive to noise, temperature and light. Extremes of any of these aspects could make a product uncomfortable, difficult or even impossible to use.

Product psychology can also influence our use of a product – it can help us to understand how the product operates. Successful products are usually straightforward, as users understand them and are confident in their use. Clever design can help the user to trust products, by making them look strong, durable or sturdy. For example, the wheels on a skateboard are chunky and, therefore, look robust.

Hint

Designers must consider how the product will make the user *feel*.

Make the Link

Adding **texture** to a product can make it more appealing to people. Sometimes it makes us want to reach out to touch products. It can also give a sense of reassurance. Textured bike handles give a feeling of added security through their grip. A textured surface can also send a subconscious signal to your brain to interact with a product, such as indentations on a button for pressing, knurling on a handle for gripping or ridges on a plastic bottle top for twisting it off.

We interact with colour coding of products every day. We know that the blue tap is cold and the red button is off. The human brain is trained subconsciously over time to automatically recognise colours.

While many products can make you feel happy or make tasks easier, overly complicated products are frustrating. Understanding how to operate a new product can be confusing, especially when it comes to technology. Complicated products can also be intimidating, making the user feel nervous about using them. For a product to be successful, it must be simple to use.

If a first aid box was pink, would it feel reliable? If a vitamin C packet was grey, would it look as if it were good for you? Would you eat this blue apple?

DESIGN

Activity

Product psychology and physiology

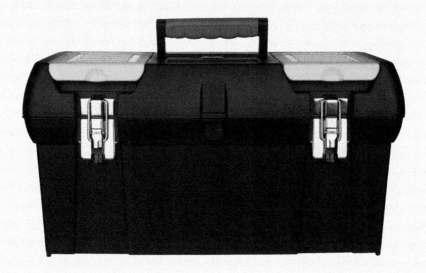

This is a paired task. The activity is to be completed on A3 or flipchart paper.

1. In pairs, brainstorm the ways in which psychology has influenced the design of the tool box in terms of:

   a) function
   b) colour
   c) texture.

2. With your partner, discuss how physiology has influenced the design of the tool box in terms of:

   a) opening
   b) closing
   c) lifting.

Health and safety

All designers have a responsibility to design products that are safe. Designers, engineers and manufacturers work with regulations that are laid down by law to ensure that we are not injured or even killed by the products we use.

Manufacturers declare that their products are safe and meet the required **standards and regulations** by labeling their products with the BSI (British Standards Institution), ISO (International Standards Organisation) and CE (Conformité Européenne or European Conformity) logos.

Products are tested to make sure that they meet the required standards. For example, a child's remote controlled helicopter must not have any mechanical or physical hazards.

The three main categories of safety breaches are strains, injuries and death.

- **Strains** – usually caused by overuse of a product. This is called a repetitive strain injury (RSI). Careful ergonomic consideration during designing should prevent strains. However, even the most ergonomically designed products can cause strains if they are overused, such as bike handles causing strains on long cycle journeys.

- **Injuries** – can mean a hospital visit for a plaster cast, stitches or even surgery. Products that cause injuries are recalled to avoid further injury or legal action. Poor material choices, badly assembled components or loose fixings can cause products to fail, resulting in an injury to the user. Not all products are to blame for injuries. A deliberately misused product can easily cause an injury, such as an adult falling off a child's swing.

- **Death** – products that contain chemicals, have blades or electrical parts are most likely to cause death. Humans have different physical abilities that impact product use, potential reactions to materials and interpretations of how to use products safely, all of which can potentially lead to fatal injuries.

Safety design considerations

When designing, avoid:

- Small removable parts, which children can swallow.
- Any sharp blades, jaggy edges or spikes.
- Finishes that are poisonous, such as toxic paint.
- Traps where your fingers can get stuck.
- Exposed electronic components or wires.
- Your discomfort; using the product should not cause you any strain.

These logos act as a reassurance to consumers and allow companies to sell their products in Europe.

Activity

Prioritising design factors

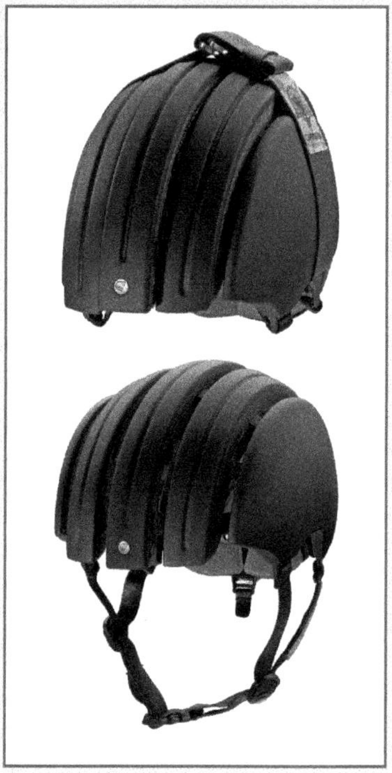

This is an individual task. The activity is to be completed in a sketchbook or jotter.

1. Research folding helmets online.
2. State the benefits that folding helmets have over standard helmets.
3. Explain how the following design factors have influenced the design of the folding helmet in terms of:
   - function
   - performance
   - market
   - aesthetics
   - ergonomics.
4. Copy and complete the diagram below. Rank the factors in order of importance, in your opinion, from the top down to the least important factor at the bottom.

Make the Link

One design factor can have more importance than the others, depending on the product.

Hint

To sustain something is to make it last as long as it possibly can. Sustainability is considering the Earth, and all its natural resources, as the most important thing to sustain.

Environmental considerations in design

Alongside the five factors that influence design, designers must also consider the environmental impact of design and how it influences society.

As designers shape our world and influence our lifestyles, they are responsible for the environmental impact of the products they design. They must ensure that products either work in harmony with our environment or positively add to it in their use and in their disposal.

Sustainability

A sustainable product is one that is designed to have minimal negative impact on the environment; it does not use finite resources, cause pollution or damage the Earth. Its manufacture takes account of the social aspects of production in that it does not damage the natural habitat to source the materials and it is not produced by people working in exploitative conditions.

> **Make the Link**
>
> Sustainable manufacturing and sustainable materials.

> **Hint**
>
> A sustainable product supports the world's sensitive natural ecosystem.

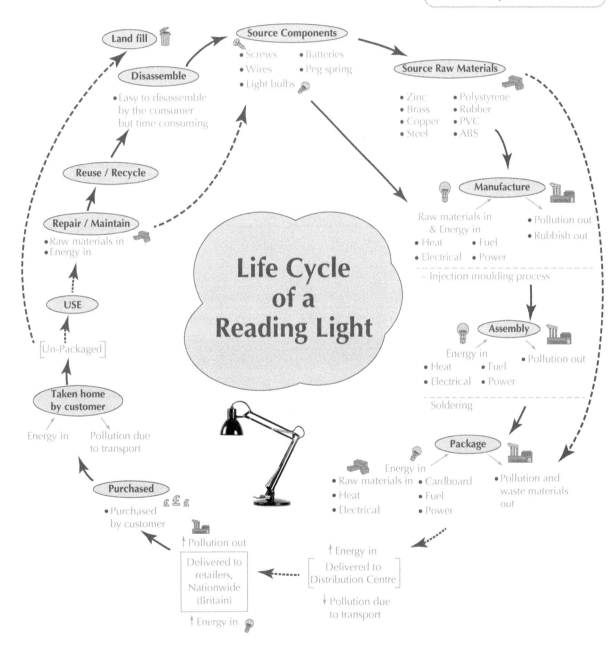

Life Cycle of a Reading Light

DESIGN

> **Make the Link**
>
> Clean manufacturing, explained in Chapter 6, requires a sustainable approach to manufacturing and the use of sustainable materials.

> **Hint**
>
> To sustain something is to make it last as long as it possibly can. Sustainability is considering the Earth and all its natural resources, so that these resources last for future generations.

There are many stages in the life cycle of a product that can impact on the environment, from the pollution caused by the transportation of materials to the factory, to the packaging of the product being disposed of in landfill.

While it is difficult to ensure that a product is made from sustainable materials, is sustainably manufactured, is sustainably packaged, is sustainably transported and is subject to sustainable disposal, efforts must be made to try to reduce the environmental impact of the product.

This coat hanger would be difficult to recycle as it is made from wood, metal and plastic.

Cradle-to-cradle design

At the end of a product's life, the sustainable approach is to recycle the product or its components. However, many products are difficult to take apart and some materials cannot be recycled. **Cradle-to-cradle design** means that products are designed to be:

- easily maintained and repaired
- made from recyclable or reusable materials and components
- disassembled.

Environmentally responsible designers consider this cradle-to-crade approach during the development of the design and products can receive certification if they meet these standards.

Check your progress

I can:

	HELP NEEDED	GETTING THERE	CONFIDENT
• explain the ways in which the function of a product influences its design	○	○	○
• explain the ways in which the performance of a product influences its design	○	○	○
• explain the ways in which the marketing of a product influences its design	○	○	○
• explain the ways in which the aesthetic of a product influences its design	○	○	○
• explain the ways in which the ergonomics of a product influences its design	○	○	○
• describe the importance of sustainability when making design decisions.	○	○	○

DESIGN

3 Designing

By the end of this chapter you should be able to:

- describe the ways in which designers identify problems
- state the purpose of a design brief
- explain different approaches to analysing a design brief
- describe research techniques
- explain a product specification
- describe the purpose of idea-generation techniques in designing
- understand ways to refine ideas in development
- recognise the common graphic techniques used by designers
- describe the modelling materials and techniques used by designers
- explain how to apply research to a design proposal
- explain how to justify design developments using the specification
- state the purpose of a sequence of operations.

The design brief

The design process diagram is on page 10.

Designing a new product usually begins with a design brief. The design brief is the first step in a series of design activities called the design process. Designers work from a design brief, whether they are designing a pair of trainers, a computer table, a hairdryer, a vacuum cleaner or a toothbrush.

Situation
We are a modern fitness centre in central Edinburgh. The fitness facilities we have are exceptional. We offer a five-star fitness experience with cutting-edge modern equipment and tailored personal training. A typical member is a young professional, who is fit and active, with a busy lifestyle.

Design Brief
Design a plastic earphone wrap which new members of Active Edinburgh would receive as a joining gift.

DESIGNING

Whether the job is an improvement to an existing product or a new invention, a designer needs a design brief to find out the details of the project and to identify any restrictions they will have to consider. The client and the design team will discuss the design brief before designing begins. Communication between the client and the design team throughout the process is the key to success. It is important to make sure there is a shared understanding of the brief.

This design brief may be short and to the point, or it may be long and very detailed. Either way, it will state what the product should do or what it should offer.

Identifying a problem

Some design briefs focus on improving current products or fixing existing problems. Many new products are simply evolved versions of products that we use every day, but are bigger, stronger, last longer or are easier to use. Therefore, a design brief is a statement of a problem, of an opportunity for a design development or of an issue that can be resolved through redesigning the product.

Case study

The Post-it note

In 1974 Art Fry worked for adhesive company 3M and in his spare time he sang in the local church choir. He had a problem, however, with his bookmark falling off his book during choir practice. He solved his problem by designing a product with light yet still quite sticky glue on a little piece of paper and the Post-it note was born.

Situation

Sometimes the designer is presented with additional information along with the design brief. This **situational information** is usually provided to give the designer an appreciation of the 'bigger picture'. The extra information may be a story about the client, details of the other products the client sells or even information about the location of use of the product. The designer must read through all the information and select the most important aspects. This information helps designers to understand the market and may stimulate ideas.

DESIGN

> ### ✔ Test your knowledge
>
> **The design brief**
>
> 1. Explain the meaning of the term 'design brief'.
> 2. Explain why good communication is important between the designer and the client when setting a design brief.
> 3. Describe a situation in which a designer has identified a problem and state the product that solved the problem.
>
> **Link to suggested answers**
> www.collins.co.uk/pages/scottish-curriculum-free-resources

> **Hint**
> To analyse a design brief and any situational information, use a highlighter to identify the key points.

Analysing a design brief

Once you receive a design brief, it is important to read it more than once. Scribble, sketch and write notes while fresh ideas and questions pop into your head. It is important not to get one idea stuck in your head and to try to let lots of ideas emerge.

In this early stage of designing, it is necessary to identify the main **design factors** that will influence the design. The design brief may include details regarding the function, performance, market, aesthetic or ergonomics.

For example, using the design brief below, a student has identified the design factors using a mindmap format.

Design brief example

Situation
People between 8 and 16 need their own space for homework, computer games and having friends over.

They are finding this space in lofts, garages or spare bedrooms, and furniture retailers are recognising the need to offer unique and specialist furniture solutions for these new living spaces. Modern, stylish products are being produced.

Design brief
Design a gaming chair for a teenager.

DESIGNING

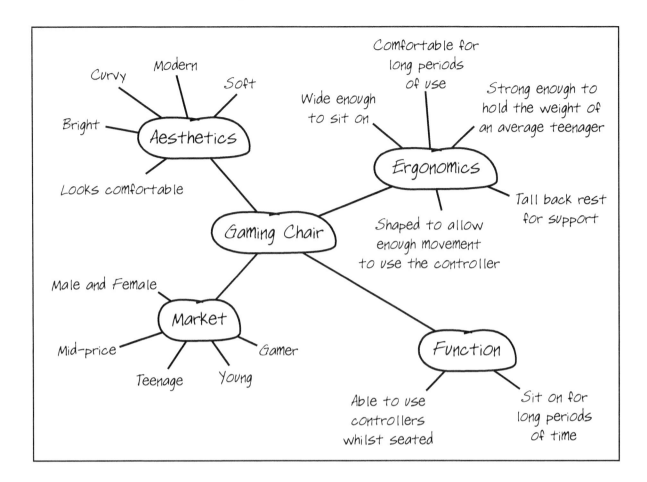

Analysis example

There are various ways to present the analysis:

- a mindmap
- written lists
- a table of information
- a recording of a discussion
- any other suitable method that shows the brief has been explored.

The analysis should identify areas to research, and you may wish to compile a list of questions raised by your analysis to use as a basis for this research.

DESIGN

> ### Activity
>
> *Analysis task*
>
>
>
> A design brief for a child's toy is given:
>
> *Design a toy for indoor use. The toy must help children under the age of three to develop fine motor skills, such as developing precise movements in their hands. It should bring learning and play together in a colourful and interesting way. The toy must be durable to endure daily use by toddlers. It must be safe to use and comply with all safety guidelines.*
>
> 1. Copy the design brief and underline the key words and phrases.
> 2. Research any words and phrases that are unfamiliar to you.
> 3. Analyse the design brief using the **five** design factors, by either drawing a mindmap or completing a table.
> 4. State the most important design factors in the design brief.
> 5. Write a list of research questions based on your analysis.

Research

Research is an important element when designing as it provides the information you need to design a successful product. Research helps designers to make meaningful design decisions and should be ongoing throughout design projects. For example, while you are developing your ideas, you may find you need to research extra ergonomic considerations that you had not thought about previously.

> ### Make the Link
>
> To get a better understanding of this ongoing approach to research, look back at the diagram of the design process on page 10.

A designer can gain valuable information about a product and they can apply this information when designing to bring improved products to the market. For example, a designer may wish to find out more about folding bicycles before they begin to design one. They would research how to operate folding bicycles, which parts wear out first, what safety risks there are, what styles or colours are popular, what size it should fold down to and so on.

Research plan

A research plan will help you to organise your work and ensure that you have selected the most relevant design factors. An example research plan is shown below.

> **Hint**
> Organise your research by design factors.

Designing a wall-mounted jewellery box for a fashionable woman in her 20s		
Design factor	**Research method**	**Research purpose**
Ergonomics	Measuring and recording	I need to find out: • What is the size of the woman's hand? • At what height should the box be mounted on the wall?
Function	Measuring and recording Using the internet/catalogues	I need to find out: • What jewellery will be stored in the box? • What standard components do I need to make my design easy to open and close?
Aesthetics	Using design magazines/blogs Survey	I need to find out: • What styles, shapes and colours are popular with the target market?

Following completion of the research plan, the designer is ready to begin their research. The questions posed will help the designer to justify the design factors they have chosen, providing reassurance to the team and the client that they are on the right track.

Research methods

When completing research, there are different approaches to gathering information.

These include:

- User trips
- Field research/visits
- Survey
- Product testing
- Measuring and recording
- Desk research – using the internet/books/magazines, etc.

Using a range of these research techniques will help to strengthen your knowledge and understanding of each method, while providing a good breadth of research material.

> **Make the Link**
> Researching products when developing design concepts.

> **Hint**
> Support your classmates by participating in their evaluations, and gain skills in evaluation methods that you may not be using yourself.

DESIGN

Carrying out research

Once a plan has been created, you should then carry out your research. Depending on your method of research, you may require:

- camera to record evidence
- measuring equipment: measuring tape, ruler, scales, protractor, etc.
- notepad/jotter/sketchbook
- equipment for cleaning/tidying up.

> **Hint**
> Before beginning the research task, visualise it from start to finish to consider the various outcomes of the research and the resources you may require.

Activity

Research plan
A child's dinner set is shown below.

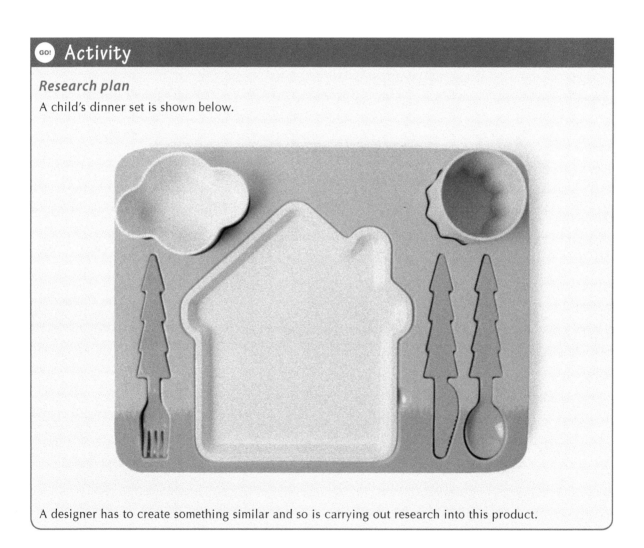

A designer has to create something similar and so is carrying out research into this product.

Activity (continued)

Copy and complete the table below:

Features	Design factor	Consideration	Questions
Has cutlery, cup and a bowl	Function		1. Can children use the cup and put it back in place? 2. Can children use the dinner set at a table and on a high chair? 3. Can the parts be easily used/removed without spillages?
Bright colours and shapes based on a child's drawing		Must appeal to young children of 12 months and over	1. 2. 3.
Easy to clean and food and drink is safe	Performance	Must be easy to clean Non-toxic material	1. 2. 3.

User trip

A user trip is when a product is operated under normal conditions of use to gain a deeper understanding of it. A user trip helps you to understand the product from the user's perspective and so it is said to be a very useful research tool for **user-centred design**, where products are designed with ergonomics, function and performance as a priority, to make products that are entirely user friendly. The user trip is a step-by-step journey through the typical use of a product, which can help you to understand how a product functions and what aspects of using it are positive or negative for the user. For example, if you are designing a vacuum cleaner, then you could carry out user trips on other vacuum cleaners to assess if the cord retracts well or the nozzle attachments are secure during use.

Key stages of a user trip

- Preparing.
- Carrying out the user trip.
- Recording and analysing the information you collected from the user trip.

DESIGN

The insulin pen is a good example of user-centred design. The pen has a digital display to make understanding it easy, while it can be discretely stored in a bag or pocket without anyone knowing it is an insulin pen.

Preparing for the user trip

Before beginning a user trip, consider:

- Who will be using the product? The age, gender, height, weight, fitness or other physical aspects.

- In which ways will the product be used? In addition to simply using the product, consider the activities that come before and after the use of the product, such as setting it up, cleaning it or storing it after use. Read the instructions carefully to use the product correctly.

- Where will the product be used? The product could be used in different locations, surfaces or weather conditions.

- When will the product be used? The user trip could be carried out at different times of the day or in certain circumstances, such as a torch being tested at night. It may require two parts to the user trip, such as one to set it up, install parts or charge it up, and then one to operate it.

DESIGNING

- Why is there a need to improve the product? What features does it have, what limitations does the product have, how could it be improved? Consider its entire journey through the user trip from set-up to storage.

Carrying out the user trip

During your research, you will need to video or photograph the user trip as evidence. You could either make written notes or give an audio commentary on the user trip video. Ideally you should carry this out yourself; however, if you are not within the target market you may need to observe the product being used by someone else. For example, the user trip for a child's trike aimed at children aged 10 months to 3 years, has to be carried out by a toddler to assess the interaction between the child and the product.

Before designing a desk, the pupil has carried out a user trip on her friend's desk. This will help her to find out about different uses and limitations of the desk, and will ultimately help her to design a desk that is an improved version of this one.

Recording and analysing user-trip data

Collating your evidence is the vital final part of a user trip, to allow you to analyse your findings.

Field research/visits

Field research or visits describe when designers go to a place outside of the design studio or workshop to gain information from a more suitable setting. If a designer is working on a chair design for a café then they should visit a few cafés to observe how chairs are used, stacked, moved around and other purposes they are used for, such as tying a dog leash or for teenagers to swing on. Field research will provide information that can only be found first hand in the location that the product will be used such as the textures, smells, activities, space and movements in the area. The **field** itself may be a house, business, hotel, dentists' surgery, shopping centre, park, gym, etc.

Many designers or engineers carry out field research to get a better understanding of how their design will fit a place or situation. For example, an architect may visit a site to see the natural flow of the land, the trees and greenery or the neighbouring buildings before embarking on a project, or a civil engineer would visit a road junction to gain an understanding of traffic flow. Field research may involve interviewing people, carrying out a survey or taking photographs at the location.

✔ Test your knowledge

Research

1. Explain the purpose of research in a design project.
2. Describe a user trip of a kettle.
3. State four different response formats for a survey.
4. Describe a test rig for a microwave.
5. Explain three benefits of field research.
6. State five ways in which research evidence can be recorded.

Survey

A survey, or questionnaire, is ideal for collecting lots of opinions about a product and for evaluating a variety of evaluation factors, depending on the product. However, some factors are difficult to evaluate through a survey, depending on the product.

Tips for a good survey:

- Mix up the format of response.
- Give the survey to the people who would typically use the product.
- Check the questions are clear before you give out your surveys.
- Give out 10 or 20 questionnaires to make adding up and working out percentages easier.
- Give the product to the person answering the survey before they begin to answer it to get valid results.
- Stay with the person whilst they answer your survey.
- Put a **return by** date on online surveys.

> **Hint**
> Conducting a survey that asks if a kitchen mop is durable enough to last 10 years, for example, would take a very long time.

To compile a survey, use a set of **closed questions** with answer options provided. You might then follow this up with an **open question**. The benefit of using a supplementary open question following a closed question is that it can provide reasons for the results and can help to justify the response.

Survey question examples

Does the textured handle make the toothbrush easier to grip?

Yes [ ]   No [ ]   ← closed question

Response: 95% said yes, 5% said no

Please explain the reason for your response: ← open question

_____

_____

Response: 30% of the responses stated that the grip benefited the user as, without it, the toothbrush would have been difficult to handle when it was wet. 60% of the responses stated that the texture allowed the thumb to grip the toothbrush. One response also stated that the height of the textured ridges made it easier to grip, compared to other toothbrushes with shallower textures.

> **Link to example Evaluation by survey**
> www.collins.co.uk/pages/scottish-curriculum-free-resources

Pictorial representations, like graphs and pie charts, are great ways to display data. But presenting written statistics can also work well.

Desk Lamp Survey

Overall the survey was taken by 26 people to answer questions about the light's aesthetics.

Q1-"What gender does the colour appeal to more?"

They were given the option to answer from 'Male, Female, or both Genders'. As the results show (shown on the right), more than half said that the colour of the light appeals to both genders. This shows that it is a unisex product and the neutral aesthetics mean that both males and females are likely to buy it.

Q2. How expensive does the material make the product look? (scale of 1-10)
Please explain your response with a comment

The average of these results is '4.81' and combined with comments such as *"the shade looks tacky"* and *"the yellow bit looks cheap"* this shows that the lamp shade makes the product look like a mid-market product. Other comments on the chrome base such as *"the chome parts look expensive"* and *"the base looks durable and sturdy and therefore makes it look high quality"* explain the responses in the high end.

Low				Question 2 Responses Mid					High
1.	2.	3.	4.	5.	6.	7.	8.	9.	10.
3.85%	0%	26.92%	7.69%	30.77%	15.38%	11.54%	0%	0%	3.85%

Q3. Does the yellow colour increase its overall appeal?

With a weighted average of 3.04, the colours minutely increase its appeal. This also shows that the colour is not overly appealing.

Not At All		Question 3 Responses Somewhat		Very Much So
1.	2.	3.	4.	5.
8.33%	16.67%	41.67%	29.17%	4.17%

Q4. If you could improve anything about the aesthetics, what would you improve?

There was an overwhelming vote for the colour to be improved, showing how much people didn't like the yellow.
Second came the shape, showing that its curved appearance and unusual form wasn't popular.
Third was materials.
Also there was an 'Other' category available for selection but no respondents chose that.

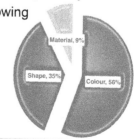

Presenting a variety of different question types will keep the person interested in the survey, so try to avoid lots of yes or no questions and mix in a range of different response options.

Try including a question with a scale of 1–10 as the answer, such as:

On a scale of 1–10, how would you rate the overall performance of the product?

1	2	3	4	5	6	7	8	9	10
Poor				Acceptable					Excellent

Another survey method is to write a statement and offer possible responses, such as:

The red colour makes the product look modern.

Agree strongly	Agree	Neither agree nor disagree	Disagree	Disagree strongly

Make the Link

A survey is also an ideal method for evaluating your product or prototype once it has been manufactured.

Simplify tricky questions by giving four options as choices, such as:

The apple slicer blades are:
a) Too sharp
b) Just sharp enough
c) Not sharp enough
d) Blunt

Make the Link

As with the user trip, the key stages of a survey are to plan, carry out and then analyse the data collected.

Many free online survey websites are available that are ideal for sharing your survey electronically, collecting responses and collating results. Some even create graphs or charts of your results for you. If you can create graphs quickly then this is a great way to display your survey results; however, simply writing the statistics can work just as well. Results should be clearly presented and an understanding of them provided.

Product testing

Product testing is another way to find out information about the way a product works. The product is put through a series of intense tests to reveal how it performs under pressure.

Here the durability of a chair is being tested in a test rig, set up to measure the force the frame can withstand before breaking.

DESIGN

> **Make the Link**
> Product testing is an ideal way of evaluating commercial products or prototypes during development.

> **Link to example Product test**
> www.collins.co.uk/pages/scottish-curriculum-free-resources

Companies often test prototypes or products using test rigs, which are set-up controlled experiments that put products through their paces at high intensity (such as extreme weights/pressures) or at a level of extreme repetition (such as being used thousands of times over and over by a a robotic arm). This method of research carries out a variety of investigations on the product. For example, repetitively tumbling the product in a metal box to test the durability of the product in extreme circumstances. Each test is measured in terms of weight, pressure or time. The results are then collated and possibly compared to results from other products. Product testing is ideal for researching similar products on the market to find flaws with their function or performance.

Product testing: function

Can I saw through the padlock?

I chose a range of metal workshop saws capable of cutting through metal.

I started with the hacksaw as it was the smallest.

It took about 30 seconds to cut through. I, therefore, did not test the other saws.

Can the padlock be forced open?

I used some pliers to try to force open the padlock, which was held in a vice.

I bent it backwards really easily with leverage against the padlock.

It popped open within about 15 seconds with hardly any force required.

Summary

The saws I chose could all cut through metal (the coping saw needed a special blade attached to it). But this would be difficult when the padlock was attached to a suitcase.

The hacksaw would be difficult to use if the padlock was not clamped in the vice and was attached to a suitcase.

The other saws could probably cut the padlock open but were not tested as the junior hacksaw proved it was possible.

The pliers would also be difficult to use if the padlock was not clamped in the vice and was attached to a suitcase. However, they would be easier to use than a junior hacksaw.

The leverage against the padlock helped to burst it open. The padlock is, therefore, not very reliable.

To record this evaluation method you can create a drawing, a storyboard of events, take photographs or make a video diary. You should also include a description of the test, the number of times it was carried out, the result of the test(s) and some notes or comments about the key facts under investigation.

> **Hint**
> Product testing is best left as the last evaluation method as a broken or damaged product can pose difficulties for other evaluation methods.

Measuring and recording

Measuring is a way to find the critical sizes that you will require to design your product. A simple example of this is to evaluate a CD rack, a CD could be measured to check that the CD could fit in the slot. This research method will help you to decide the critical sizes. Research should be ongoing throughout the development of the design. It may be necessary to carry out some additional measuring and recording during the development stages. You can draw simple diagrams annotated with measurements during development to help refine the size of your product.

> **Make the Link**
>
> If you plan to use standard components in your design, then you will need to measure and record these components.

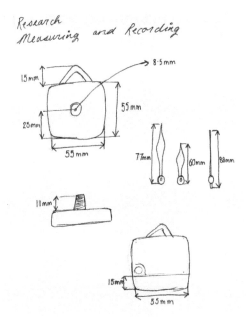

As part of a student's clock design, they have measured the components for a clock and recorded them clearly.

Desk research

Desk research is using the internet, books, magazines, catalogues, etc. to source information. The internet is a great tool to find out information; however, it can be difficult to find exactly what you are looking for if you have specific ideas. Use different search criteria to locate the facts such as: hinge sizes, sizes of standard hinges or what size is a butterfly hinge?

Using books is also a great way to source information. This includes using books/catalogues that your school technician may use to order components, materials or parts. They contain lots of details and often contain good quality graphics of items.

Design magazines are useful to source up-to-date fashion trends, while in consumer magazines you can find product comparisons and expert advice.

Additonal reading material such as receipts, packaging, instructions, guarantees and marketing material (adverts, flyers, etc.) can often be overlooked, but may include important information.

> **Link to example Evaluation by measuring**
>
> www.collins.co.uk/pages/scottish-curriculum-free-resources

> **Hint**
>
> Record the websites you use as you go. Keep a note of them in a jotter or sketchbook, add them to your favourites or use a pin-board style website that allows you to access this information from different computers.

Collating and analysing research

When completing research, you will gather a range of evidence. When presenting your research findings, a clear layout is important. This helps to make your work easy to understand and use when you are designing.

Your research evidence may include:

- A collection of images
- Written information
- Annotated sketches
- Labelled diagrams
- Tables of information
- A series of photographs
- Video evidence

Your research should provide the answers to the questions or statements you compiled for your research plan.

Whatever research method you use, it is important to write up results clearly showing your evidence and outcome. Here are some phrases you may wish to use:

- The survey set out to determine ...
- In this part of the research, the aim was to ...
- One of the more significant findings to emerge is ...
- This result proves that ...
- Extensive testing of the product has shown that ...
- The findings of the user trip indicate that ...
- Surprisingly the product was found to be ...
- As expected the results prove that ...
- Generally the majority agreed with ...
- The following conclusions can be drawn from this ...
- Without doubt this confirms that ...
- The most obvious finding to emerge from this survey is ...
- Analysing the information revealed ...
- The evidence suggests that ...
- When these results were examined they revealed that ...
- This evidence verifies that the ...

> **Hint**
>
> In addition to recording and analysing your research, you may record the key research findings at the end of each research activity or as a summary of all your research if you have a report format for your research.

Product specification

A **specification** is a list of things that the product must do. It gives details about the product with precise information and facts. The specification is written following discussions with the client and analysis of the design brief, and is based on the outcomes of any research.

Designers use specifications as checklists to ensure their designs match the design brief and to avoid designing something unsuitable. The specification should be presented as a list of clearly defined statements that correspond to the research gathered. No information should be added in at this stage, unless it is recorded as part of the research.

When designing, it is a good idea to keep your specification handy. Designing within the specification helps designers to ensure that they do not overlook or forget about any of the main points.

> **Hint**
> A design that does not meet the specification will not meet the brief.

Write the specification points on Post-it notes and stick them next to where you are working.

Presenting the product specification as a list

This first method of presenting a specification requires an opening statement followed by bullet points.

The soap dish must:

- hold a bar of soap with a maximum size of 70 × 50 × 30mm
- have a durable waterproof finish that will not rot in a humid environment
- be available in different finishes or materials to appeal to a wide target market
- be made from a material that will not rot in a damp environment
- be shaped to hold a bar of soap
- be easy to clean with water
- be waterproof
- be easy to assemble
- be designed with a modern aesthetic.

> **Hint**
> The key word in a specification is 'must'.

Presenting the product specification by design factors

The second method of presenting a specification is to list the points by design factor.

DESIGN

Function
- The soap dish must hold a bar of soap of dimensions 70 × 50 × 30mm.
- It must be shaped to hold a bar of soap.

Durability
- The soap dish must have a durable waterproof finish or be made from a waterproof material.
- It must be made from a material that is suitable for a damp environment.

Aesthetics
- It must be available in different finishes or materials to appeal to a wide target market.
- It must be designed in a modern style.

Performance
- The soap dish must be easy to clean with water.
- It must be waterproof so that it does not come apart when in contact with water.
- The soap dish must be easy to assemble.

> **Make the Link**
> Later on in the design process you can revisit your specification to update it.

Activity

Product specification

A designer receives the following design brief for a bookcase:
A school library wishes to display a selection of recommended books that will be updated on a weekly basis. The bookcase should be able to store books of different sizes. The librarian has chosen a central location for the bookcase, intending that it will become a focal point in the library, attracting the attention of lots of pupils. Preferably, the bookcase should be able to be accessed from the front and the back, so it has to be sturdy and free standing. It will be used by a range of pupils aged 11 to 18, as well as by adults such as the librarian and teachers.
The library itself is located centrally in the school, which is in a new building. It is a bright, modern and spacious place with areas for quiet study and reading.

1. Write a product specification for the bookcase; produce a list with **10** bullet points.
2. Write a product specification for the bookcase by listing the points by design factor, using the following **five** design factors: function, performance, market, aesthetics and ergonomics.
3. State which method you prefer to use when presenting a product specification.
4. Explain why you find this method easier.

Idea-generation techniques

Designers often use idea-generation techniques to stimulate their creative thinking.

There is a range of idea-generation techniques and there is no particular technique to use in any specific situation; the choice of technique simply depends on the client, the project and the design team. Some designers have their own preferences. However, the technique *must* foster new ideas.

When you develop design ideas, you must show how you have used your selected idea-generation technique to find inspiration. You should justify its purpose in your designing. For example, a student might write a simple comment next to one of their sketches that states, *"I used the analogy idea-generation technique – the shape of the spout on this design is inspired by the natural fold of a leaf."*

> **Make the Link**
>
> Look at the spout on the watering can on page 74, which demonstrates the use of analogy in design.

Brainstorming

This technique works best with a group of people as more ideas can be created. However, it can also work with small groups or pairs. Brainstorming is an opportunity for the team to share their ideas and see the project through the eyes of other group members. The team can build on and improve ideas as they are suggested. Brainstorming should have a fast and productive pace.

A team leader records all the responses and manages the group. A good team leader will ensure everyone participates and all the ideas are recorded without editing or question, as long as they relate to the topic. The results can be written as a list, diagrams, written as notes pinned to a notice board or can even be doodles or sketches. A successful brainstorming session will be a supportive teamwork exercise with only positive comments. It will include a range of fun, obvious, unusual, boring, weird and wonderful ideas.

> **Make the Link**
>
> As with the research methods, the key stages of an idea-generation technique are to plan, carry out and then analyse the data collected.

A marketing team brainstorm a marketing strategy to generate new ideas for the colour and style of the product.

DESIGN

> **Hint**
>
> The word 'analogy' means to compare two things that are similar to help to describe something unfamiliar. An analogy includes the term 'is like'. For example, *'life is like a box of chocolates. You never know what you're gonna get.'*

Analogy

Analogy is used to find a new train of thought by thinking about similar products, comparable circumstances or by looking at nature.

The easiest way to approach an analogy is to write a list of important features of the design and then consider these questions:

- What other products does it remind you of?
- Where have you experienced something similar?
- What is the product like?

Bio-mimicry is a type of analogy that uses nature for inspiration. It can be a powerful and imaginative approach to generate ideas. One example of a product that was designed using bio-mimicry is Velcro. It takes its inspiration from a burdock plant that has burrs (small hooks) on the surface of its seeds. In nature, the hooks are used to disperse seeds as they catch onto an animal's fur or a person's clothing and the seeds fall off in a new location. Velcro uses similar hook technology – these latch onto a softer surface, creating a reusable joining method.

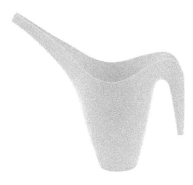

The design team working on this watering can investigated water in nature to discover the ways water flows naturally. Then, when designing the shape of the spout, they shaped it with the idea of water pouring over a folded leaf.

Mood boards

A **mood board** is a collage of images that relate to the project. Designers use mood boards to collect their ideas and to inspire new ideas. Mood boards are popular with graphic designers, product designers, interior designers and architects as they help to create an aesthetic vision for the project ahead. A mood board will investigate the patterns, textures, shapes and colours the product could have.

This mood board aims to generate ideas for a Christmas cake stand. The images have a crisp and sophisticated feel, which can inform aspects of the plate design.

Lifestyle boards

Designers use **lifestyle boards** to understand the person who will use the product and, therefore, to gain a deeper understanding of the target market. A lifestyle board looks just like a mood board. However, it has different content. The images in a lifestyle board represent the lifestyle of the typical person who would use the product.

This technique helps the designer to understand the user by considering where they live, what they eat, what they wear, what hobbies they have, what music they prefer, what pet they have, where they go on holiday, etc.

A lifestyle board for a female consumer in her twenties.

Morphological analysis

The technique of morphological analysis uses the information that you have collated so far (from the design brief and specification for example) to create this set of categories, such as material, colour, style and manufacturing method, and then listing possible responses for each category.

These categories and lists can then be entered onto a grid (or matrix), so that entries can be combined randomly and can be used as inspiration for a design idea. The purpose of this technique is to create unusual combinations of attributes that the designer may not have considered previously; aiding their creativity, extending their thoughts into new areas and, therefore, broadening the range of ideas.

One alternative method of creating a morphological analysis is to collate your lists on strips of paper, cut a letter-box style slot in a piece of card and place the strips behind the card. Move the lists up and down to create different word combinations, which are viewed in the slot. Another simple method is to circle randomly one word in each column on your matrix and then combine these to generate ideas.

> **Link to example Morphological analysis**
> www.collins.co.uk/pages/scottish-curriculum-free-resources

Morphological analysis of a lamp			
Location	**Feature**	**Market**	**Style**
Dining room	Touch control	Child	Rock and Roll
Bedroom	Uplight	Adult	Circus
Office	Dimmer	Elderly	Seaside
Garage	Adjustable	Baby	Formal
Classroom	Spot	Teenager	Sporty

Technology transfer

Transferring technology from existing products into new design ideas can bring very interesting results. It involves considering the design properties that are top priority and thinking about other products that can meet these priorities. The analysis involves making a list of these priority features and entering against each feature the name of a product that can deliver that feature.

A successful technology transfer session will:

- list the features required in the design
- list other products that have these features
- incorporate a method of researching products during the process.

DESIGNING

Case study

The Dyson roller ball

James Dyson designed the Ballbarrow, a wheelbarrow design with technology transfer. The plastic wheel was designed like a ball; spherical so it was easy to roll, hollow so it was lightweight, and plastic so that it could be used outside. The Ballbarrow was, therefore, easy to manoeuvre on soft ground, thanks to the large surface area of the ball.

In 2005 Dyson then went on to transfer this technology into his design for the ball-based upright vacuum cleaner, which is now a familiar product in homes around the world. Just like the Ballbarrow, the vacuum cleaner is noted for its excellent manoeuvrability.

Application of idea-generation techniques

Designers use these techniques throughout the design process and some ideas might be useful for small details, not an entire design project. Idea-generation techniques can be used not only to create design concepts or develop ideas, but throughout the design process. For example, a designer could brainstorm the most efficient/cheapest/quickest ways to manufacture a product with a manufacturer. This would help them to write a plan for manufacture. Similarly, a morphological analysis could be used to come up with different scenarios (location, user, conditions of use, etc.) to evaluate a product.

Hint

It is important not to sit looking at blank paper. During the next few stages of the design process, remember to use idea-generation techniques if you find you have run out of ideas.

Test your knowledge

Link to suggested answers
www.collins.co.uk/pages/scottish-curriculum-free-resources

Idea-generation techniques

1. Describe the purpose of idea-generation techniques.
2. Name three idea-generation techniques.
3. Explain **one** way in which brainstorming can encourage creative thoughts.
4. State the name of the idea-generation technique where the designer looks for a similar occurrence in nature to generate ideas.
5. Name one example of technology transfer.
6. Describe how morphological analysis could help a designer to improve the breadth of their ideas.

DESIGN

Link to example Initial ideas
www.collins.co.uk/pages/scottish-curriculum-free-resources

> **Hint**
> Don't forget to have the specification next to you as you design. Use it as a checklist to ensure your designs stay on track.

> **Hint**
> Creating a page of quickly-sketched ideas or using a sketchbook to continually draw design ideas will take the pressure off putting pencil to paper.

Design ideas

With the information gathered from the design brief, analysis, research and any idea-generation activities carried out so far, designers have established a good understanding of the project and they can begin to create design ideas, or concepts.

Designers will begin to explore ideas through freehand sketching, basic diagrams, written notes/annotations and sketch modelling. Indeed, designers may have made several doodles already to help them visualise their ideas.

To communicate design ideas, the graphics and models must be clear and detailed where appropriate. Designers will make use of a range of graphic and modelling techniques to present a breadth of designs that match the specification and have scope for development.

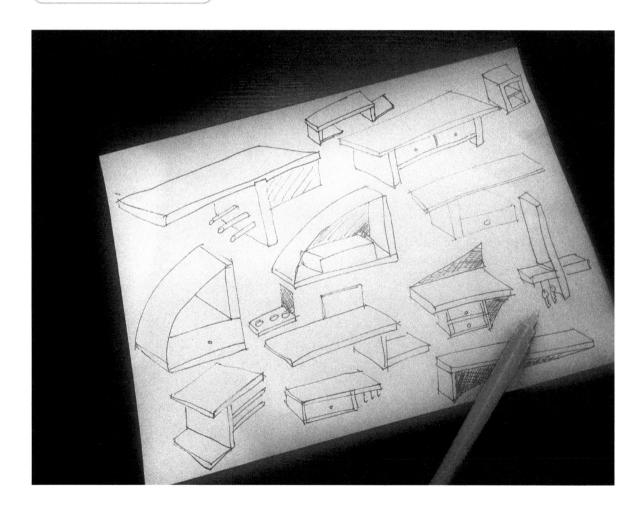

Using graphic and modelling techniques throughout the design process

During the design process, the style of the graphics techniques applied changes:

- Design ideas – presented using 2D and 3D sketches with simple diagrams, some sketch modelling and potentially even some basic/rough computer-generated models. Due to the sketchy nature of these methods, ideas can flow more naturally to the designer.

- Design development – presented using 2D and 3D sketches and drawings with additional diagrams, 3D modelling and computer-generated models. Due to the development taking place, some aspects may be quickly sketched whilst other areas are drawn, proportioned and rendered to a higher degree of quality.

- Design proposal – highly detailed 2D and 3D drawings showing the full aesthetic and technical qualities of the final design.

A student has produced a range of freehand sketches to communicate design ideas.

> **Link to example Synthesis**
> www.collins.co.uk/pages/scottish-curriculum-free-resources

Design development

When a number of different design ideas have been produced, designers then **refine** or **synthesise** these ideas to give one design to develop further.

So, **synthesis** is the process of evaluating and judging the design ideas. This can involve morphing some of the best features from a few different designs into one stronger design idea.

> **Hint**
> Synthesis can be approached as a single stand-alone task or can be undertaken as a series of justifications during the development.

DESIGN

> **Hint**
>
> When beginning the development stage, start with a statement that explains which design is to be developed and give reasons why.

During this stage, all the ideas that are considered unsuitable are discarded. The designs that are most feasible and most appealing are developed.

The next task is to **develop** this design idea by:

- producing a range of sketches, drawings and diagrams using different graphic techniques
- writing notes
- annotating your work
- creating a range of models.

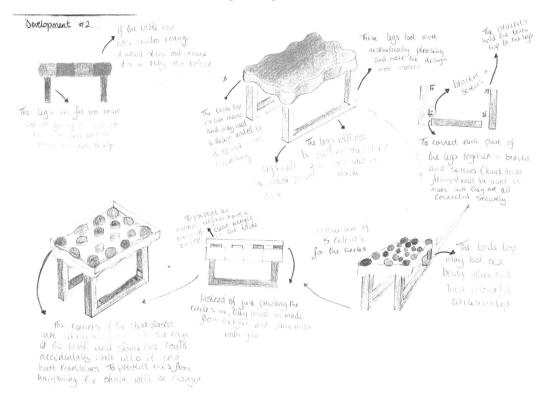

A range of graphic techniques can be used when developing design ideas.

> **Link to example Development**
> www.collins.co.uk/pages/scottish-curriculum-free-resources

Graphic techniques to use in development

When designing, a range of graphic techniques should be used to generate and explore ideas in 2D and 3D. These techniques include:

- isometric (to show the 3D form of design ideas)
- oblique (to show the 3D form of design ideas)
- perspective drawings (to show the 3D form of design ideas in realistic perspective views)
- working drawings (to show the dimensions, detailed sizes, scale and joining methods of design ideas)

DESIGNING

- exploded drawings (to show how component parts fit together)

- computer-generated models (to show all of the above)

> **Make the Link**
>
> Proportion is also explained in the section on Aesthetics (see page 39).

> **Link to example**
> **Exploded view**
> www.collins.co.uk/pages/scottish-curriculum-free-resources

A racing car manufacturer uses a working drawing to visualise the design.

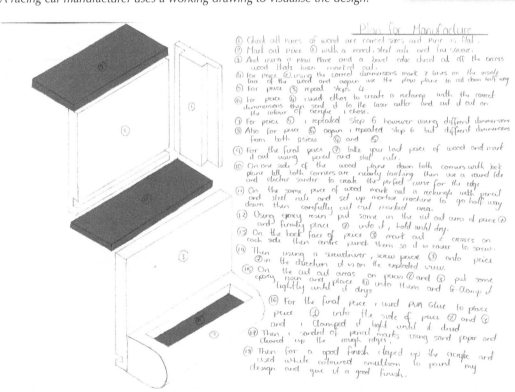

Designers consider which graphic technique will be most suitable to communicate the development of ideas. For example, if the drawing is to show how the product is assembled, then an **exploded view** may be the most suitable method of communicating this.

Designers use **scale** to create a true representation of their designs. Drawings appear more realistic if the **proportion** of the design is accurate.

> **Hint**
>
> Including a familiar object (such as a coin or pencil) in a scale drawing can further communicate the scale in context. The object must also be drawn to scale.

A basic guide for drawing to scale:

Size of product	Examples	Scale
Small products	Kitchen utensil	1:1
	Desk tidy	
	Candle holder	
Mid-sized products	Picture frame	1:1 or 1:2
	Key cabinet	
	Magazine rack	
Larger products	Bookcase	1:4 or 1:10
	Coat rack	
	Bird table	

To draw to a scale of 1:2, divide all the sizes by two. For example, if the height of a picture frame is 240mm, then the drawing will be 120-mm high.

When drawing large products, consider the scale before putting pencil to paper. To draw to a scale of 1:4, divide all the sizes by four. For example, if the width of the product is 160mm, then the drawing will be 40-mm wide.

A design for a lamp has been sketched in context to show its scale and proportion.

DESIGNING

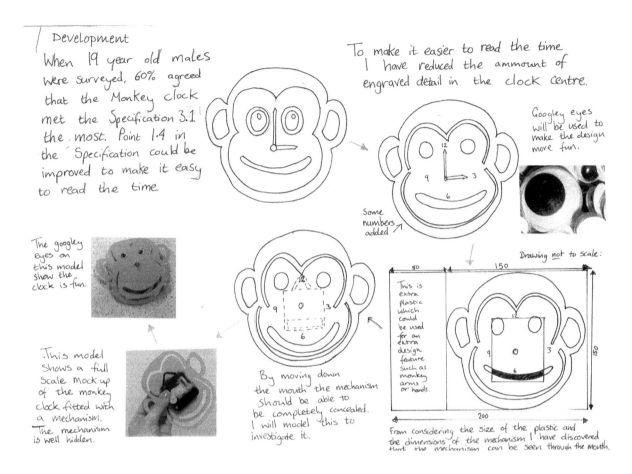

Development at National 4 and National 5 can incorporate a range of graphic and modelling techniques. Research can also be ongoing during development, such as measuring standard components.

Applying colour and texture **rendering** will bring your ideas to life. This can be done with a range of media:

- pencils (2B–6B are ideal for rendering)
- coloured pencils
- watercolour pencils
- pastels
- marker pens
- CAD rendering.

Adding texture will help to communicate the materials and convey the tactile qualities of the design. Good texture rendering will enhance the overall quality of the drawing, making it look more realistic.

> **Make the Link**
>
> Use the Colour theory section on page 41 to help you select colours that suit the product in terms of function, ergonomics and aesthetics.

DESIGN

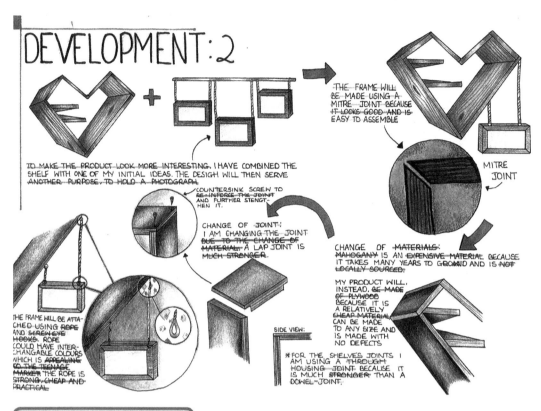

Link to example Rendering work
www.collins.co.uk/pages/scottish-curriculum-free-resources

A pupil has used pencils to create a realistic wood effect on their development page.

Activity

Texture rendering

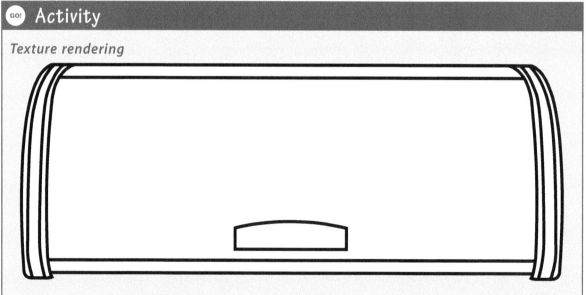

1. Trace the bread bin and add a wooden texture to it using coloured pencils or marker pens.
2. Trace the bread bin again and add a metal texture to it using marker pens or graphite pencils.
3. Trace the bread bin again and add a plastic texture to it using pastels or marker pens.
4. Sketch or draw your own bread bin and render it with a medium of your choice.

Modelling techniques

Models can communicate ideas in 3D, in addition to supplying drawings and sketches or as an *alternative* to drawings and sketches. The purpose of modelling is to help visualise shapes, forms and proportions in three dimensions. The process of creating and refining a 3D model can provide opportunities to generate new ideas and can also be used to solve problems.

A student has made a block model of a storage unit to evaluate the size of the product on the wall.

Modelling materials vary from things you find in your recycling bin to specialist modelling materials, such as balsa wood. Some popular modelling materials are:

- paper
- card
- corrugated card
- MDF
- wire
- pipe cleaners
- foam
- clay
- modelling compound
- balsa wood
- expanded foam
- sheet plastic
- construction kit materials
- smart materials.

> **Hint**
> Consider the purpose of the model before selecting a type of model.

Using models to generate and explore ideas

Models can be made at any point in the design process; while generating ideas, problem solving or developing designs. It is not essential always to model the entire design – some models may be produced to show specific details, such as a joining method or a component part. Creating models is not just an activity that is used at the end of the design process to show off the final design, but is an approach to visualising design solutions throughout the process.

DESIGN

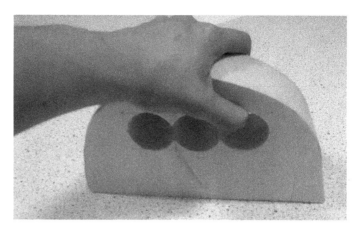

A full-scale model made from modelling foam is used to test ergonomics.

Make the Link
Models can be used to communicate a final design to a client in a design proposal, but they are also useful in other ways.

Link to example Modelling work
www.collins.co.uk/pages/scottish-curriculum-free-resources

Advantages of using modelling in design

The purpose of modelling is to:

- **m**easure and test the success of the design in terms of ergonomics, aesthetics, etc.
- **o**pen a discussion with clients to get feedback or present a final proposal
- **d**etermine how parts fit together
- **e**valuate the design and any problem-solving solutions
- **l**earn about the strengths and weaknesses of a design.

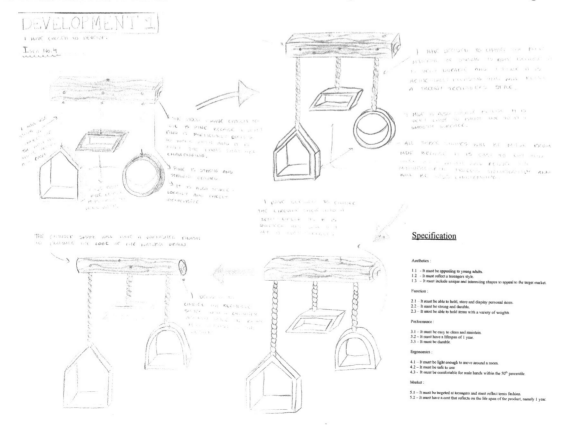

Test and refine models

When producing a model, it is important that the type of model used suits the intended purpose in order for it to be tested and improved. For example, when beginning to develop your ideas, a range of mock-up models could be made from clay to show different shape possibilities in 3D. Then, as development continues, a wood scale model could be used to investigate proportion and size. As the development is refined, a computer-generated model could be produced to explore exact sizes and to show how the component parts fit together.

Communicating the results of modelling

When using models to develop your ideas, you should explain the purpose of the model. A simple annotation to describe the purpose *and* outcome of the modelling process will make a valuable addition to your work. Try starting your annotations like this:

- This model shows …
- Through modelling this design I discovered …
- When I produced this model I decided …
- This model was used to test …

DESIGN

✓ Test your knowledge

Communicating Ideas in 2D and 3D

While developing a vanity unit, the designer manufactured a scale model from balsa wood.

> **Link to suggested answers**
> www.collins.co.uk/pages/scottish-curriculum-free-resources

1. Explain the benefit of a 3D model when showing the vanity unit to the client.
2. State the name of an alternative modelling material for the scale model of the vanity unit.
3. State the name of another model type the designer could create and explain its purpose.

The designer used a range of graphic techniques during the design development.

4. State the names of **three** graphic techniques the designer could use to communicate ideas in 3D.
5. State the name of **one** graphic technique the designer could use to communicate ideas in 2D.
6. Describe why texture rendering is important when designing.

Annotations and notes

Designers use annotations and notes to clearly communicate their design decisions

Annotation: a short piece of information that is tagged onto a drawing. Annotations can be one or two words, dimensions (sizes), arrows (for example to show opening/closing mechanisms), symbols, calculations, sub headings and generally anything short that adds detail to the design.

> 🔍 **Hint**
> Continually review your ideas to show an ongoing evaluation of your development.

Note: a sentence that is written next to a drawing to provide more in-depth explanations than annotations. Notes are ideal for justifying design decisions. For example: 'The cookie cutter is made from aluminium, which will not rust'. When writing notes, be careful not to write too much. Consider that small diagrams, models or sketches can also explain technical information and take up less space (and time!).

Justification

During designing, you must be able to **justify** your design decisions, especially when evaluating and developing your design ideas. This requires some designerly thinking: applying all your knowledge gained from the research to develop your ideas to reach a suitable design.

These justifications can be communicated through drawings, sketches, models, annotations or notes.

Justification of your design development will involve:

- incorporating the findings of your research
- referring to your idea-generation techniques
- including references to the design specification
- referring to the design factors
- discussing the suitability of materials and manufacturing techniques
- considering the sustainability of the design and its impact on the environment.

Finalising the development

Once a final design has been reached, it may require some final tweaks, further investigation or detailing to ensure it is the best possible solution. This may involve making and testing a prototype, or creating a computer-generated model to work out the sizes.

Some questions to ask yourself at this stage are:

- Do I know which materials will be used for all the parts?
- Are these the best materials for the job?
- Do I know the method of manufacture for all the parts?
- Have I selected the most suitable manufacturing methods for the materials?
- Do I know how the parts will join together?
- Am I sure there are no better ways to join these parts?
- Have I considered the sizes of all the parts?
- Can I make this design more environmentally friendly?

Hint

A drawing of the design proposal is just one part of the information you need to progress to the next stage.

Development checklist

- Sketches or drawings in 2D and 3D using graphic techniques.
- Consideration of scale, proportions and dimensions.
- Clear improvements to the design.
- Justification of design decisions.
- References to the specification, design factors and research.
- Notes and annotations.
- Colour and texture rendering.
- Consideration of manufacturing methods and suitable materials.

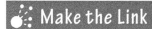

Make the Link

Earlier in the design process a basic, less detailed specification was written. Now that you have more facts about the product, a more detailed specification can be written.

DESIGN

Presentation Drawing

Design proposal

When a final design is reached, this must be presented or proposed to the client. This is an opportunity to explain the design decisions and justify the final design.

Presentation of the final design proposal

Designers can produce a **presentation drawing** as their main method of communicating the final design proposal. This may be a computer-generated drawing or a manual illustration of the product. It must show the final design proposal in a way that is easy to understand. Another way to communicate a final design proposal may be to make a fully functioning **prototype** and take photographs of the main features of the design. These photographs could be annotated to explain their relevance to the specification and design factors.

Sequence of operations

This is a set of instructions for how to manufacture the product you have designed. For this you will need knowledge of the materials, tools, machines, joining methods and surface finishes.

> **Hint**
> Look ahead to the next chapters for information about materials, tools, machines, joining methods and surface finishes.

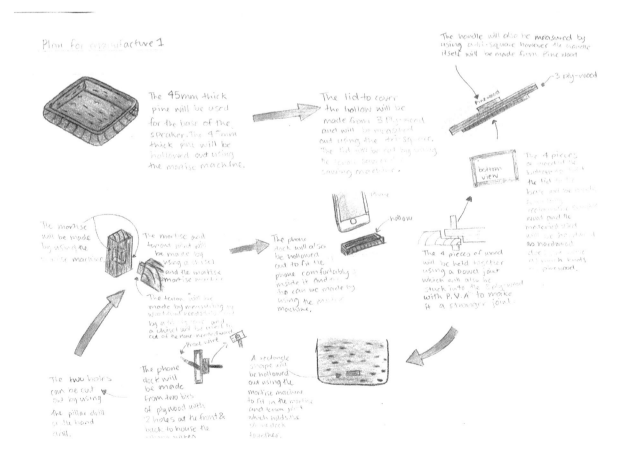

DESIGNING

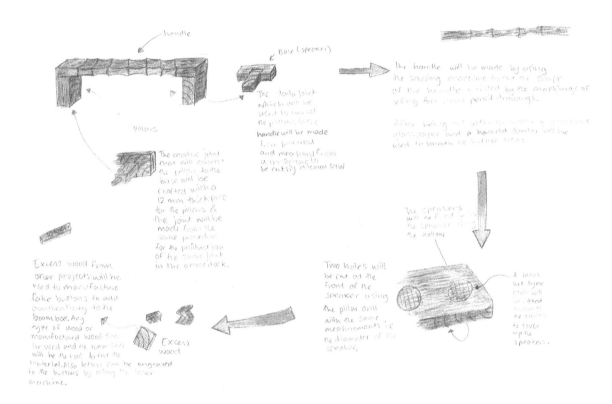

Manufacturing steps can be communicated with ease when drawings are provided, even simple 2D sketches or diagrams are useful.

A good way to approach the plan for manufacture is to write a list of the manufacturing tasks you will need to carry out. Next, discuss your list with your teacher to check the tasks are in the correct order and you haven't missed any stages. Then produce a more detailed list of instructions for each stage to explain the manufacturing techniques, assembly methods, sizes, tools required, finishing techniques and any other relevant details. This information, together with **cutting lists** and **component part lists**, should explain the manufacture of the product.

> **Hint**
>
> Don't underestimate the time it takes to achieve a good finish. Set aside a few hours at the end for finishing and remember that varnish or paint may need more than one coat.

Working drawings

During manufacturing, a **technical drawing** is often required to find sizes, details of parts and joining methods. This is called a working drawing. It usually includes an **orthographic drawing** (manual or computer generated) with dimensions. However, a range of 3D drawings with dimensions can also be used.

DESIGN

> **Hint**
> Use a working drawing when you are in the workshop.

The standards for orthographic drawings are

- drawings are well-proportioned
- an appropriate scale is used
- 3rd angle projection layout (plan is directly above the elevation, end elevation is directly in line with the elevation)
    - dimensions are placed to the side of the drawings, not over them
    - avoid repeating dimensions
    - limit the number of views to a minimum
    - dimensions are all in mm
    - construction lines, heavy outlines, centre lines and hidden detail line types are used
    - drawings and diagrams to show all the essential information and details.

In industry, working drawings are very detailed, with many drawings produced for each product. These drawings are given to all the production workers and they must be accurate to avoid any mistakes. Incorrect working drawings can cost the factory time and money.

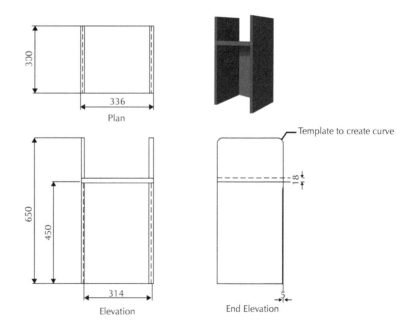

A working drawing has been created for a bedside table to be manufactured from MDF.

DESIGNING

Cutting list

The cutting list is part of the plan for manufacture. It is a table of information that contains the materials and parts required for the product. Manufacturers use cutting lists to order materials before the manufacturing begins and they refer to it during manufacturing to check sizes.

> **Link to example Working drawing**
> www.collins.co.uk/pages/scottish-curriculum-free-resources

A good cutting list will include:

- part names – keep them simple (left side, base, etc.)
- length – the length of the material
- breadth – the breadth of the material
- thickness – the thickness of the material
- sizes in millimetres
- quantity – number of parts required.

Cutting list

Part	Material	Length	Breadth	Thickness	Quantity
Bottom	Pine	500	300	16	1
Top	Pine	350	300	16	1
Back	Pine	648	300	16	1
Feet	Pine	45	45	20	4
Support	Plastic rod – red	648	200 ø	-	2
Support	Plastic rod – orange	648	200 ø	-	2
Support	Plastic rod – green	648	200 ø	-	1

Preparing for manufacture

At this point in the design process, the product should be ready for manufacture.

Check that you have:

- a sequence of operations
- a working drawing
- a cutting list.

You can now prepare for manufacture by cutting materials to size, checking there is enough paint, screws, adhesive, etc. and buying any additional parts you need.

Hint
Use your working drawing to work out sizes.

DESIGN

Activity

Cutting list

Part name	Material	Length	Breadth	Thickness	Quantity
Shelf	Pine		50	12	1
Rail	Pine			20	
Pegs	Pine				
Mirror					

1. Copy the cutting-list table above.
2. Complete the cutting list for the hall storage unit using the information from the working drawing below.

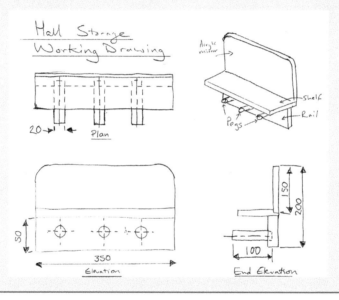

Evaluation

The purpose of the evaluation is to investigate the success of the product once it has been manufactured. Research methods can also be used to evaluate your product. These include:

– comparison to the specification

– comparison to similar products

– user trial

– survey/questionnaires

– product testing.

Comparison to the specification

To evaluate the success of your project, you can compare your finished prototype to your specification. This should help you to

DESIGNING

determine a basic indication of the areas where the product has been successful and where it has not. For example:

A chalkboard has been created for a homeware shop.

Specification statement: The chalkboard must have space to write the shop's deals on it.

Evaluation: The board is narrow but it still has enough space to write a few words or a line of text.

Specification statement: The chalkboard should fit inside the porch of the shop.

Evaluation: The width of the chalkboard is 500mm, therefore it will fit inside the 650-mm deep porch area.

Further evaluation will help validate these points:

While the chalkboard is narrow and can only fit a few words of text on each line, it has to fit in the porch for storage. Therefore, it reaches the happy medium of being a reasonable size while being suitably sized for storage.

Comparison to similar products

Comparing your prototype to other similar products on the market is ideal for evaluating the prototype. To carry out a comparison to other products, first identify which design factor/factors you wish to evaluate using this method. Then gather the same information about other similar products.

> **Link to example Product evaluation**
> www.collins.co.uk/pages/scottish-curriculum-free-resources

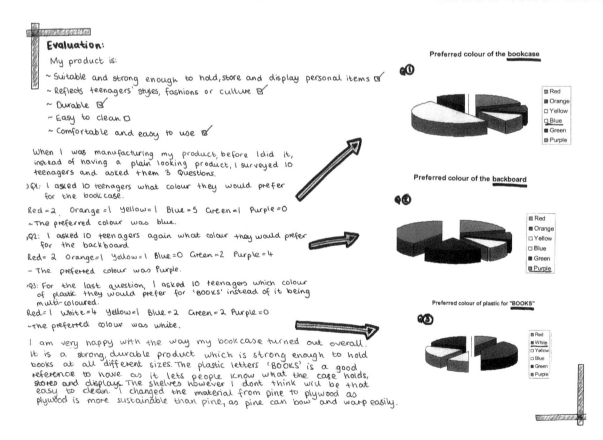

Check your progress

I can:

	HELP NEEDED	GETTING THERE	CONFIDENT
describe the ways in which designers identify a problem	◯	◯	◯
state the purpose of a design brief	◯	◯	◯
explain different approaches to analysing a design brief	◯	◯	◯
describe research techniques	◯	◯	◯
explain a product specification	◯	◯	◯
describe the purpose of idea-generation techniques in designing	◯	◯	◯
understand ways to refine ideas in development	◯	◯	◯
recognise the common graphic techniques used by designers	◯	◯	◯
describe the modelling materials and techniques used by designers	◯	◯	◯
explain how to apply research to a design proposal	◯	◯	◯
explain how to justify design developments using the specification	◯	◯	◯
state the purpose of a sequence of operations.	◯	◯	◯

CONTENTS

- **An introduction to materials**
- **Manufacturing in the workshop**
- **Commercial manufacturing**

MATERIALS AND MANUFACTURING SECTION OVERVIEW

The second section of this book will provide you with the knowledge and understanding for **Unit 2, Materials and manufacturing**. It will also help you to complete your course assignment. Combined with your classwork, workshop experience and the guidance of your teacher, these chapters will help you to develop the knowledge and understanding required to manufacture your own prototype. At the end of this section, there are also some examples of the types of questions you might meet in the National 5 question paper.

Chapter 4, **An introduction to materials**, contains useful information on different materials, their unique properties, identifying features and uses. Sustainability issues associated with materials throughout the life cycle of a product are described. The aim is to make you aware of your responsibility as a designer to use materials wisely. Also in this chapter, you will learn how to justify why a particular material is appropriate for a manufacturing task.

Chapter 5, **Manufacturing in the workshop**, explains common manufacturing methods used in a school workshop. The tools, machinery and equipment used to manufacture with metal, plastic, wood and manufactured boards are listed. This chapter aims to help you to select suitable manufacturing methods for a component or product. The information in this chapter will be useful to your study for Section A of the National 5 question paper. It will also be useful for the course assignment.

Chapter 6, **Commercial manufacturing**, explains the methods by which commercial products are manufactured in factories. Computer aided manufacture features in this chapter and the commercial manufacturing processes for wood, metal and plastic products are described. The drawbacks and benefits of these processes are explained. You will gain an insight into the global impact of commercial manufacture and the associated sustainability issues. This is all essential knowledge and understanding for Section B of the National 5 question paper.

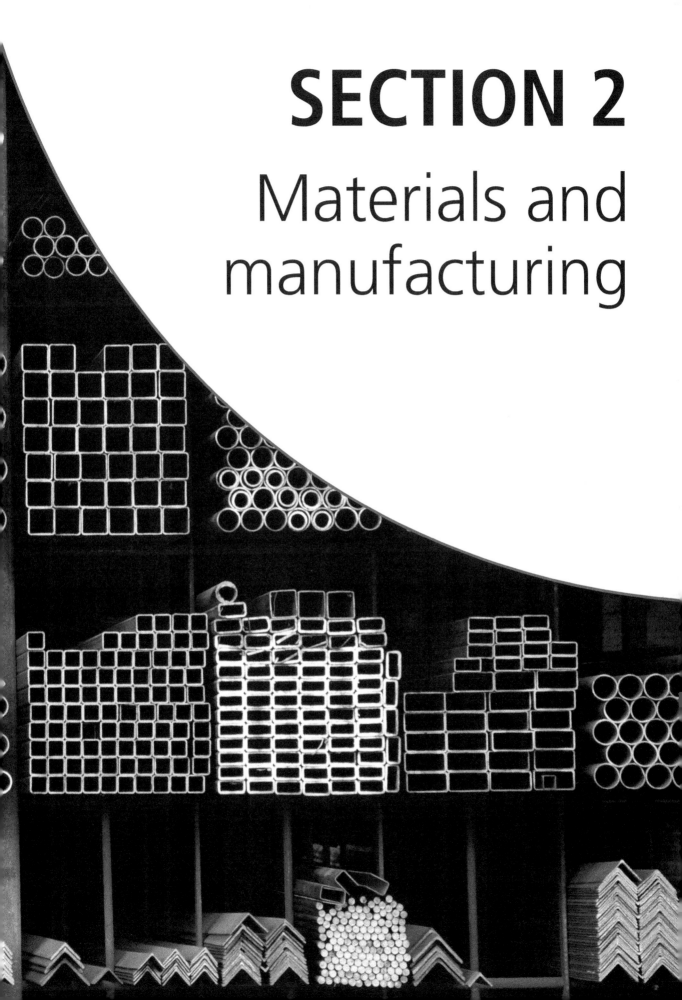

SECTION 2
Materials and manufacturing

MATERIALS AND MANUFACTURING

4 An introduction to materials

> **By the end of this chapter you should be able to:**
> - describe the sustainability issues related to materials
> - describe the properties and identifying features of a range of woods and manufactured boards
> - describe the properties and identifying features of a range of metals
> - describe the properties and identifying features of a range of plastics
> - explain reasons why a material is suitable for the manufacture of a component or product
> - describe the sustainability issues related to materials.

Selecting a material

When selecting a material for a product, there are lots of considerations, such as:

- availability
- properties
- environmental impact and sustainability issues
- conditions of use of the product
- suitable manufacturing processes
- cost of the materials
- source.

The properties of the material must match the needs of the product. For example, a wood with a high natural oil content should be used for outdoor furniture as it will last longer outside in bad weather.

The volume of products to be produced should also be considered. For example, using an entire board of MDF to make one small key rack will waste a lot of material. It would be less wasteful to use a material that is supplied in smaller sizes, such as a plank of pine.

Clever material use is an environmental consideration.

> **Hint**
> Check which materials are available before making any decisions, to avoid disappointment.

AN INTRODUCTION TO MATERIALS

Sometimes, choosing a material depends upon the manufacturing process or joining method. For example, chipboard can be difficult to work; it may not be possible to cut a good dovetail joint with this material. Using a softwood such as spruce wold be much easier. To avoid potential problems, take the demands of the manufacturing process into account before selecting a material.

> **Hint**
> When selecting a material, designers should try to find a local supplier to cut down pollution from transportation.

Properties

Materials have special characteristics and features, or **properties**. Properties can include, for example, being strong, soft, brittle, shiny, waterproof, lightweight, malleable, durable, and many others.

Once a material has been selected, it must be tested to check that the properties match the requirements of the product. If the material isn't suitable for some reason, then another material should be selected. You must be able to explain why you have recommended a particular material, using the evidence you gathered when testing it.

Copper is used for central heating pipes because it has suitable properties: it is easy to bend, is corrosion resistant and keeps its shape even at high pressure and high temperature.

Forms

Materials are available in a variety of forms. The **form** is the shape of the material when it is supplied. Examples of forms include sheet, bar, plank, tube, rod, wire, etc.

Uses

The term 'uses' refers to the products or purpose for which the material is used. Aluminium is a suitable material choice for a drinks can as it is lightweight, does not rust and can be recycled.

Identifying features

The visual characteristics of a material are its **identifying features**. These are the elements of materials that make them recognisable and they have a big impact on the aesthetic of the product. For example, the identifying feature of copper is the distinctive reddish-brown colour, and a glossy shine is characteristic of acrylic. There are many features of a material that make it identifiable, including the colour, pattern, weight, texture and even the way it responds to testing, such as burning.

The neutral and plain hardwood used for this button-shaped pot stand has no wood knots and so the other aesthetic aspects such as shape and style stand out.

Standard sizes

Materials are supplied in standard sizes. These are the common sizes of boards, planks, rods, sheets, etc. When designing, it is important to consider standard sizes and to design with them in mind. For example, dowel rod is available in 6, 12 or 15-mm

MATERIALS AND MANUFACTURING

> **Make the Link**
>
> Chapter 6 explains the ways in which products can have identifying features as a result of processes during commercial manufacture, such as ejector pin marks left by injection moulding.

diameters; to design a product with 13-mm diameter dowel rod would not make sense.

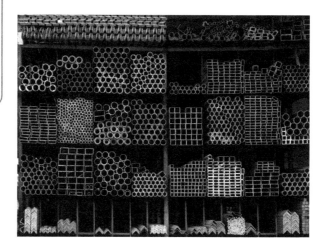

A metal store stocks metal in standard lengths and widths.

Wood

Choosing a type of wood demands careful consideration as there are lots of varieties. Different types of wood have different strengths, visual features and colours. There are two main categories of wood: **hardwood** and **softwood**. Additionally, a third material, which is made from timber, is **manufactured boards**.

Wood is a natural material and, therefore, each piece is unique. The fine lines on the wood are called the **grain**. Sanding in the direction of the grain will smooth the surface of the wood. Sanding across the grain will scratch it.

Wood also has knots – dark circles on the wood where a branch once grew. Knots are tough and can be difficult to work with but look very interesting and give a unique aesthetic quality to the material.

Some wood, such as walnut, is used for its unique grain and vibrant colours.

Wood properties

The properties of wood and manufactured boards vary depending on the type of material; each wood and board has its own unique set of properties. The properties of wood can include:

- Buoyant – able to float.
- Durability – the resilience of the material (perhaps when outdoors or with wear).
- Density – the combined weight, strength and durability, ranging from soft to hard.
- Natural oil content – natural oil provides natural weatherproofing.

- Porous – the grain is open and has small 'pores', making it absorbent.
- Stable – flat and smooth, so doesn't warp or twist. Stable materials will paint or stain well.
- Strength – the ability to withstand force without breaking, ranging from weak to very strong.
- Toughness – the ability to withstand sudden blows or shocks without breaking.
- Weight – the heaviness, ranging from lightweight to heavy. When a material has a good strength-to-weight ratio then it is strong for a lightweight material.
- Workability – the ease of cutting, shaping, etc.; ranging from easy to work to difficult to work. Usually, if wood is difficult to work it is due to the grain or knots. Manufactured boards can be difficult to work due to the way they are made.

When selecting a wood for a product involving lots of cutting and shaping, avoid woods that are known for having lots of knots as they are difficult to work with.

Wood supply

The common sizes of timber materials are:

- **Manufactured boards** – available as a standard size of 2440 × 1220mm or 1220 × 607mm with a thickness of 6, 9, 12, 16, 19, 22mm.
- **Planks** – available in various sizes depending on the type of wood.
- **Strips of wood** – available in various sizes depending on the type of wood.
- **Square battens** – available in 35 × 35mm, 45 × 45mm, 75 × 75mm, 100 × 100mm.
- **Dowel rods** – available in diameters of 6, 8, 10, 12, 15 and 25mm.

Advantages of using wood

Designers use wood as it has a unique aesthetic and it is considered to be a comforting and warm material. Wood products have a handcrafted feel, adding a sense of skill, quality and care. Using a wood with an interesting or unique grain pattern brings a distinctive style that no other material can match. As trees can be replanted, wood is a sustainable material. Wood can also be recycled and reusing it is a fashionable way of sourcing material.

● MATERIALS AND MANUFACTURING

Disadvantages of using wood

As wood is a natural material, it typically requires a finish to protect and waterproof it. Applying a finish adds cost and time to manufacturing. Under pressure, wood can split along the grain and loose wood knots can fall out. Using wood can also mean working with unwanted natural characteristics, such as bowing, splitting, warping and twisting, which can add cost and time to manufacturing.

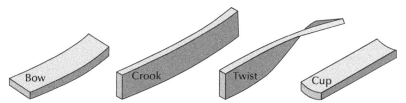

At the sawmill, wood is seasoned to dry it out. However, irregular grain and tension within the wood can cause it to bow, split, warp or twist. Storing wood at the wrong temperature or moisture level can also cause these defects.

Hardwoods

Hardwoods usually come from deciduous trees, which are trees with leaves, not needles. Hardwood trees have leaves and can be identified by their leafy branches in summer. In winter, most shed their leaves.

Not all hardwoods are hard.

Despite their name, hardwoods are not all hard. The lightest wood in the world, balsa wood, is a hardwood. The term 'hardwood' comes from their higher resistance to water, compared to softwoods, making them less likely to rot. As they are slower to grow than softwoods, hardwoods are generally more durable, but also more expensive.

Hardwood trees lose their leaves over the winter months.

AN INTRODUCTION TO MATERIALS

Name	Properties	Identifying features	Uses	Forms
Beech	Strong and durable but easy to work with and finish.	Pale brown colour, distinctive flecks in the grain.	Toys, tool handles, furniture (used for steam bending).	Plank, square batten, strip.
Mahogany	Moderately strong, easy to work with, durable and finishes well with oil, wax and varnish particularly.	Light brown to reddish brown with a close and even grain, which gives a smooth and even surface.	Furniture, shop fittings, bar tops, solid wood flooring.	Plank, square batten, strip, veneers (wide planks are also available due to the large size of trees).
Oak	Strong, tough and durable but very difficult to work with.	Open grained with distinctive markings. Looks expensive.	Furniture, building beams, barrels, solid wood flooring.	Plank, square batten, strip, veneers.
Ash	Strong, tough and durable with good elasticity (suitable for bending with steam).	Pale brown with a straight grain, which can have a coarse texture.	Furniture, hammer handles, garden tool handles, hockey sticks, boat oars.	Plank, square batten, strip, veneer.

Hardwoods can be identified by their colourful grain patterns and lack of wood knots.

> ### 🔍 Hint
>
> The properties of a material can be linked directly to its use, for example beech is a strong wood so it is suitable for tool handles. Birch, walnut, teak and balsa are other hardwoods that your school may have for you to use.

Test your knowledge

Hardwoods

1. Outline three ways in which a hardwood tree can be identified.
2. Name a dark-coloured hardwood.
3. Describe three identifying features of beech.
4. Explain two uses of ash. You must refer to the material properties in your answer.
5. Explain why hardwoods can be used as a veneer.

Softwoods

Softwoods farming is a sustainable approach to forestry, creating a constant supply of cost-effective, good-quality timber.

Softwoods come from coniferous trees, which are trees with needles and cones (e.g. pine cones) instead of leaves. Softwood trees can be identified as being evergreen all year round. The tree trunk tends to have a lot of branches growing from it, therefore there are more wood knots in a softwood than in a hardwood.

Softwoods are quicker to grow than hardwoods and so they usually cost less than hardwoods. They are ideal for sustainable forestry (in which a new tree is planted when another is cut down), because they are fast growing. Approximately 80% of woods used are softwoods. They grow in Scandinavia, northern Europe, Russia and North America.

Softwood trees keep their needles over winter and can endure low temperatures.

AN INTRODUCTION TO MATERIALS

Name	Properties	Identifying features	Uses	Forms
Red pine	Strong, straight-grained but difficult to work due to knots.	White to pale yellow colour, straight grain and lots of brown coloured knots.	Used for DIY, furniture, construction work and simple joinery.	Plank, square batten, strip.
Spruce	Strong, resistant to splitting and easy to work.	Pale white, small knots and sometimes contains resin pockets.	Used for building timber, furniture and musical instruments (pianos, violins, guitars).	Plank, square batten, strip.

Douglas fir, cedar and Scots pine are other softwoods that your school may have for you to use.

Test your knowledge

Softwoods

1. Outline three ways in which a softwood tree can be identified.
2. Name a pale white-coloured softwood.
3. Describe three identifying features of red pine.
4. Explain two uses of spruce. You must refer to the material properties in your answer.
5. Explain why softwoods are ideal for sustainable forestry.

Manufactured boards

These boards are made from wood pulp, blocks, chips or strips. They are mostly inexpensive compared to solid wood as they are made from wood that would normally go to waste. The main benefit of manufactured boards is that they are available in large sheets, whereas the size of solid wood planks is limited by the tree trunk diameter. Manufactured boards are said to be stable, meaning they have a flat and even surface, which is ideal for a smooth finish. Also they do not warp, twist or bow like timber.

Manufactured boards do however contain adhesives, so they are dangerous to use, particularly MDF. Inhaling a lot of MDF dust can cause very serious health problems, even cancer. Some types of manufactured boards are not very durable, such as chipboard furniture as it crumbles over time at the edges. It is common for manufactured boards to be laminated (covered with a thin layer of hard-wearing plastic such as Melamine Formaldehyde) or veneered (covered with a thin layer of wood). This adds to the cost of the material. To use manufactured boards outside is not advisable as they do not weather well.

 Hint

Red pine gets its name because the bark on the tree is reddish brown. The wood itself isn't red.

Larger trees can provide wider planks. However, they take longer to grow and so are quite expensive.

MATERIALS AND MANUFACTURING

Name	Properties	Identifying features	Uses	Forms
Plywood	Stable and durable. Strong due to alternate grain direction in each layer.	Thin layers of wood, each layer placed at 90° to each other.	Furniture, cabinets, worktops, building construction and under flooring.	Boards.
Flexi-Ply	Stable and durable. Able to bend into curved shapes.	Thin layers of wood, each layer is laminated to allow the board to flex.	Furniture, lighting, small wooden products.	Boards.
Chipboard	Prone to splitting when used with screws. Difficult to cut and shape. Edges chip easily.	Compressed wood chips with a solid wood veneer on the top.	Worktops, cabinets, desk tops, low-cost furniture and shelving.	Boards. Laminated boards (with a plastic top layer) are also available.
MDF	Stable, stiff, easy to cut and shape. MDF dust is extremely harmful to breathe.	Wood fibres pressed together, satin smooth surface texture, rougher edges.	Furniture, worktops, cabinets, desk tops, low-cost furniture and shelving.	Boards. Laminated and veneered MDF boards are also available.
Hardboard	Very soft with little strength, bends easily and stable on one side only. Thin boards stay flat but cannot hold weight.	Wood fibres pressed together. Top is smooth and the back is soft and almost furry.	Templates for workshop use in manufacturing, picture-frame backs, wardrobe/bookcase back panels.	Thin boards (3 and 6-mm thick only).

Veneered chipboard is a common material used in flat-packed furniture. Edges without veneer are covered by other panels through clever joins or are back facing.

Case study

Yellow Broom

Yellow Broom is a small, two-person business based within the Cairngorms National Park in Scotland. They make handcrafted lighting with a strong emphasis on design, quality materials, good craftsmanship, function and strong aesthetics. Their designs are primarily made from wood and feature softwoods shaped into elegant forms, combined with the modern twist of flexi-ply. Inspired by the local environment of the Cairngorm Mountains, Yellow Broom have a clear state of mind, which helps them to create work with strength and simplicity.

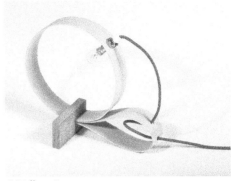

© Yellow Broom

Yellow Broom value sustainability and are proud to create products with a zero-waste approach to production. All excess materials from off-cuts through to bends that have snapped and experiments that have not succeeded are carefully reused; some are converted into wooden business cards, the rest are collected and carefully stored to create one-off lights. They present themselves with the challenge of redesigning their waste to create a unique one-off product range from what otherwise would be discarded.

The shavings from the turning process carefully provide the packaging, while all other Yellow Broom waste can go into the log burner that provides the workshop with heat over the long, harsh Highland winters.

Test your knowledge

Manufactured boards

1. Explain three benefits of using a manufactured board.
2. Explain three drawbacks of using a manufactured board.
3. Outline one way in that plywood can be identified.
4. Name a manufactured board that is smooth on both surfaces but has rougher edges.
5. Describe three identifying features of hardboard.
6. Explain two uses of flexi-ply. You must refer to the material properties in your answer.
7. Explain why chipboard is often veneered.

MATERIALS AND MANUFACTURING

Make the Link

The tables on pages 112 and 113 show the properties of each metal.

Metal

Metal ores are rocks that are mined and then processed to refine them into metals. Some metals are mined as pure metals, such as copper, while some are mixed together to improve their properties, such as brass (copper with added zinc). A metal made from a mixture of two or more metals is called an **alloy**. All metals, including alloys, are classified into two main groups: **ferrous** and **non-ferrous**.

Metal properties

The properties of metal vary depending on the type of material; each metal has its own unique set of properties. The properties of metal can include:

- Brittle – cracks or breaks easily.
- Corrosion resistant – does not rust.
- Ductility – can be stretched out without breaking to make long, thin lengths of metal.
- Durability – the resilience of the material and its resistance to wear and tear.
- Thermal insulator – able to maintain temperatures.
- Electrical conductor – able to pass an electric current through the metal.
- Malleable – soft enough to be shaped by pressing or pushing without the metal breaking or springing back into shape.
- Strength – ability to withstand force without breaking, ranging from weak to very strong.
- Toughness – the ability to withstand sudden blows or shocks without breaking.

Advantages of using metal

Metals come in a range of colours – from greys to golden yellows. Painting or plastic dip-coating metal gives even more scope for colour choice. Without a finish, metals can be polished to a high shine, which makes products appear glossy and new; therefore, metal products are said to have a hygienic and sterile appearance. Metals are excellent for conducting heat and electricity, and some metals, such as aluminium, have a good strength-to-weight ratio.

Disadvantages of using metal

Ferrous metals are prone to rusting and so a finish must be applied to them, adding time and cost to manufacturing. Once a finish has been applied, it will require maintenance to ensure it does not chip or flake off and further coats may be required, perhaps on an annual basis.

Although all metals can be recycled, metals are a finite resource, meaning that once all the metals have been mined from the Earth, they cannot be replaced. Therefore, metal products should be designed so that they are easy to disassemble, reuse or recycle.

Metal supply

When selecting a metal, there is a wider range of forms and standard sizes compared to wood and plastic. The most common sizes of metals are:

- **Wire** – available in a range of thicknesses and lengths.
- **Bar** – metal bar is available in a range of cross sections (the shape at the end of the bar) such as round, square, flat (rectangular) and hexagonal. These are available in varying sizes from 5mm to 50mm. Lengths of bar are 1000 or 2000mm.
- **Pipe** – 5-mm to 40-mm diameter. Lengths of pipe are 1000 or 2000mm.
- **Sheet** – metal sheets can come in extremely thin sheets, just 0·1-mm thick, up to 10-mm thick. However, thicker metals are also available.
- **Ingots** – large blocks of metal that are melted down to use for moulding metals.
- **Powder** – fine metal powder is used for 3D printing.

Ferrous metals

A ferrous metal is a metal that contains iron. It can therefore rust, as iron reacts to oxygen in the air and in water and develops iron oxide (rust) on the surface of the metal. To avoid rusting, ferrous metals require a finish, such as paint, to protect the metal. Ferrous metals can be identified with a simple magnet test, as ferrous metals are magnetic.

Wrought iron can be forged and bent, whereas cast iron is heated until it becomes molten metal and is then poured into a mould.

MATERIALS AND MANUFACTURING

Stainless steel is well known for being corrosion resistant, so is an excellent choice for a sink.

Name	Properties	Identifying features	Uses	Forms
Iron	Strong, tough, cannot be bent or forged. A hard outer skin.	Usually painted to avoid rust, very dark grey to black with a matt texture.	Baths, garden benches, pots, weights, drain covers, railings machines.	Ingots, bar and pipe.
Mild steel	Ductile and malleable yet tough. Easy to weld.	Silvery grey with poor corrosion resistance.	Car bodies, nuts, bolts, nails, screws, wire fencing.	Ingots, wire, bar, pipe, sheet.
High-carbon steel	Brittle, difficult to cut, resistant to wear.	Silvery grey with poor corrosion resistance.	Tools: saw blades, files, screwdrivers chisels, etc.	Ingots, wire, bar, sheet.
Stainless steel	Corrosion resistant, tough, difficult to cut and shape.	Silvery grey. Can appear dull or shiny.	Sinks, cutlery, saucepans, furniture, lighting, bikes, kitchen appliances.	Ingots, wire, bar, pipe, sheet.

✔ Test your knowledge

Ferrous metals

1. Explain a simple test used to identify a ferrous metal.
2. Explain why ferrous metals rust.
3. Name a suitable ferrous metal for car bodies.
4. Describe two identifying features of iron.
5. Explain two uses of stainless steel. You must refer to the material properties in your answer.

Non-ferrous metals

A non-ferrous metal is a metal that does not contain iron. It is, therefore, not magnetic. As they do not contain iron, most non-ferrous metals are corrosion resistant, which means they do not rust. Non-ferrous metals can tarnish (dark-coloured spots or smears appear on the surface) but this can be polished off most metals and it only occurs after many years.

Copper tarnishes green with exposure to oxygen.

Name	Properties	Identifying features	Uses	Forms
Aluminium (pure metal)	Good strength-to-weight ratio, malleable, conducts heat and electricity well, ductile.	Shiny silvery grey colour that polishes well.	Kitchen foil, drinks cans, boat hulls, sports equipment.	Ingots, wire, bar, pipe, sheet.
Brass	Rigid, polishes well, conducts heat and electricity well, solders well.	Golden tones that polish well.	Taps, plaques, ornaments, instruments, household fittings.	Ingots, wire, bar, pipe, sheet.
Copper	Tough, ductile, malleable, conducts heat and electricity well, solders well.	Deep orange brown colour but tarnishes green with exposure to oxygen.	Electric wires, pipes, circuit boards, jewellery, roofing, cooking pans.	Ingots, wire, bar, pipe, sheet.

Aluminium is an unusual material in that it keeps its original properties when it is recycled. It does not downgrade when recycled and so is ideal for drinks cans.

Test your knowledge

Non-ferrous metals

1. Explain a simple test used to identify a non-ferrous metal.
2. Explain why non-ferrous metals do not rust.
3. Name a suitable non-ferrous metal for sports equipment.
4. Describe two identifying features of brass.
5. Explain two uses of copper. You must refer to the material properties in your answer.
6. Explain why recycled aluminium is as good as new aluminium.

Make the Link

The tables on pages 116 and 117 show the properties of each plastic.

Plastic

Plastic is a synthetic material. The basic raw materials used in the manufacture of plastics are made from oil, natural gas and coal. Plastics are classified into two main groups: **thermoplastics** and **thermoset plastics**.

Properties of plastic

The properties of plastic vary depending on the type of material; every plastic has its own unique set of properties.

The properties of plastic can include:

- Buoyant – able to float.
- Brittle – cracks or breaks easily. Non-brittle plastics are shatter resistant.
- Chemical resistant – does not wear away or break down when exposed to chemicals.
- Durability – the resilience of the material, resistance to wear and scratching.
- Flexible – how bendy the plastic is; from soft to rigid.
- Good heat insulator – able to maintain temperatures by keeping cold items cold and hot items hot.
- Heat resistant – able to withstand high temperatures without deforming, burning or melting.
- Moisture resistant – the ability to repel liquids.
- Strength – ability to withstand force without breaking.
- Toughness – the ability to withstand sudden blows or shocks without breaking.

Plastic supply

When selecting a plastic, the supply types and sizes are complex.

- **Sheets** – available as a standard size of 1220 × 607mm, with a thickness of 4mm.
- **Rods** – available in diameters of 3 to 100mm. Lengths vary.
- **Tubes** – available in sizes of 5 to 100mm. Lengths vary.
- **Powder** – available for mass manufacturing processes in large volumes.
- **Granules** – available for mass manufacturing processes in large volumes.
- **Synthetic fabrics** – a standard width of 150mm and available in any length.

Plastic granules used for industrial manufacturing processes.

Advantages of plastic

Plastic is a very useful material as it is generally very lightweight but strong, and therefore it has a good strength-to-weight ratio. There is no need to apply a finish to plastic as it is waterproof and has built-in colour. It can also be transparent, making it the only alternative to glass. Plastics can resist chemicals in a way that wood and metal cannot. They have a smooth surface that can be wiped clean and so are said to be hygienic.

Acrylic comes in a range of colours and transparencies.

Disadvantages of plastic

Some plastics can be tricky to cut and shape as they can easily crack or shatter, and plastic dust is extremely dangerous to inhale. The surface of some plastics can also scratch easily (this is why most sheet plastics are supplied with a thin protective film on them).

Over time, plastics can become weaker with wear and sunlight can bleach the colour. Plastics cannot withstand extreme temperatures. They will deform, blister or burn if they are exposed to heat, while low temperatures can actually make plastics more brittle. Petroleum-derived and non-bio plastics can have an extremely negative environmental impact if they end up in landfill.

These laundry Pegs will bleach in the sun over a period of time.

Thermoset plastics

When they are heated and shaped, thermoset plastics set. They cannot be returned to their original shape by reheating them. This means that they can withstand higher temperatures than thermoplastics and are used for products that are exposed to heat, such as oven dishes, electric plugs and sockets. Due to their ability to withstand heat, they are extremely difficult to

MATERIALS AND MANUFACTURING

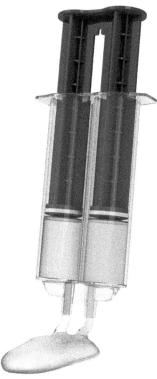

Epoxy resin is another thermoset plastic. It can be used as an adhesive and comes in two tubes or a double pack like the one shown here. When the two tubes are mixed together the chemical reaction causes the liquid to set and become a hard plastic mass.

recycle. Recycling a thermoset plastic involves grinding it down into a powder or small chips, which uses a lot of energy and the grainy end result has very few uses. It is currently debatable whether the energy required to recycle a thermoset plastic makes it worth recycling it at all. Repurposing the thermoset plastic components, such as upcycling parts for other products, may be better for the environment.

Thermoset plastics can withstand high temperatures.

Name	Properties	Identifying features	Uses	Forms
Melamine formaldehyde	Rigid, tough, scratch resistant, heat resistant.	Grainy texture.	Table tops, work tops, cooking utensils.	Veneers
Urea formaldehyde	Rigid, strong, heat resistant.	Matt or glossy finish.	Adhesive, electrical plugs and sockets.	Resin, laminates, fabrics.

✓ Test your knowledge

Thermoset plastics

1. Explain why thermoset plastics can withstand higher temperatures.
2. Explain why oven dishes are made from thermoset plastics.
3. Name a suitable thermoset plastic for electrical sockets.
4. Explain two uses of melamine formaldehyde. You must refer to the material properties in your answer.
5. Explain why recycling a thermoset plastic may not be a sustainable approach.

Thermoplastics

Thermoplastics soften when heated, allowing them to be shaped. They harden into shape as they cool. With this type of plastic, the softening and hardening can be repeated many times over. **Re**heating a thermoplastic will **re**turn it to its original shape, unless it has been permanently damaged by excessive heat or deformation. This unique characteristic of thermoplastics on reheating and reshaping is known as plastic memory (i.e. the material **re**members its original shape).

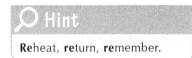

Reheat, return, remember.

Name	Properties	Identifying features	Uses	Forms
Acrylic	Rigid, very durable, polishes to a high shine.	Glossy and smooth texture.	Shop signs, picture frames.	Sheet, rod, tube, granules, powder.
Polystyrene	Lightweight, rigid, moisture resistant.	Glossy and smooth. Can be transparent.	Model kits, packaging, CD cases, plastic cutlery, toys.	Sheet, rod, tube, granules, powder.
Polypropylene	Lightweight, flexible, resists cracking and tearing.	Waxy surface, often transparent or semi-transparent.	Ice cream tubs, straws, kettles, climbing ropes, crisp packets.	Sheet, rod, tube, fibres, granules, powder.
ABS	Very tough, scratch resistant, chemical resistant.	Shiny surface, colourfast.	Casings for electronics, car body parts, toys, luggage.	Sheet, rod, tube, granules, powder.

✔ Test your knowledge

Thermoplastics

1. Explain two reasons why thermoplastics plastics can be recycled.
2. Explain why climbing ropes are made from polypropylene.
3. Name a suitable thermoplastic plastic for luggage.
4. Explain two uses of acrylic. You must refer to the material properties in your answer.
5. Explain why shaping a thermoplastic product using the strip heater will create less waste if mistakes are made.

MATERIALS AND MANUFACTURING

Researching materials

As part of researching during designing, it is a good idea to test some materials to check their suitability for a design before they are used in manufacture. This is important, as the material influences the success of the product.

This can take place as part of the research stage or as ongoing research throughout the design development. Something simple such as testing how easy red pine is to saw using a coping saw or to find out if a piece of acrylic can be bent to a particular shape in your design using an oven and former. Even experimenting with small material samples can provide enough information to make informed decisions. By carrying out simple tests, appropriate materials can be identified and you can then write a good justification for your material selection.

You can also supplement this by carrying out some desk research to find out how sustainable the material is and get some advice from your teacher as to the cost and whether the material is available for you.

A pupil is sawing a piece of red pine with a coping saw to research the ease of creating a curved side panel for his design.

Here are some key phrases to help when writing about materials:
- (The material) cut well with good quality, clean cuts.
- (The material) was harder to cut than (another material) but had better quality cuts.
- (The material) was smooth and so when I painted it, the finish was good.
- (The material) has a lovely natural colour that shone when it was cleaned with steel wool.
- (The material) was very difficult to file in a vice.
- (The material) was tricky to clamp due to its shape.
- (The material) was easily dented.
- After some time researching (the material), I noticed a number of scratches on it from general workshop tasks.
- The structure of (the material) made it difficult to work with.
- When drilling (the material), the drill bit cut through easily.
- The wood knots on (the material) were difficult to work with.
- (The material) was easy to cut but fell apart at the edges.
- (The material) required time to heat and soften them before folding.
- (The material) did not require heat to fold it.
- (The material) bubbled if left on the heat for too long.
- (The material) could be reshaped if mistakes were made.
- (The material) cracked slightly along some of the folds.

Recommending final selection of materials

It is crucial to justify material choices in designing. A designer will clearly describe the reasons for the material selection by identifying the properties of the material along with any research to validate their choice. The reasons for choosing a material can be based on:

- Aesthetic – qualities of how a material looks naturally and whether a finish is required.
- Sustainability – whether they can be recycled, locally sourced or environmentally sourced.
- Market – how available the materials are to the manufacturer and how affordable they will be to the target market.

- Ergonomics – if the materials meet the needs of the end user (weight, texture for grip etc.).
- Performance – how long the material will last or other performance needs.

Here are some examples:

The rich, reddish brown colour of the mahogany makes it suitable for the spice rack as it works well with the traditional style of the kitchen and complements the orange, red and brown tones of the spices.

The cookie cutter will be made from aluminium because it was malleable and easy to shape, it did not rust in the dishwasher test and it was found to have a sharp enough edge to cut the dough.

After testing a few different woods on the wood lathe, it was decided that beech would be the best material to make the candle holder. The red pine proved difficult to turn due to the number of knots. The close grain of the beech made it easy to work and suits the design aesthetically.

Along with the properties of the material, the tests should identify an appropriate material.

Sustainable materials

From sourcing a raw material from nature, to the end of a product's life, every material has an impact on the environment. A sustainable material can be regrown or replaced, can be recycled and will not harm the environment.

Finite resources

Sourcing materials to manufacture products can cause environmental damage. Mining materials from the Earth, such as ore for metal, must be done responsibly as these are **finite** or **non-renewable resources**. Over-mining can ruin landscapes, affect wildlife by causing loss of habitats and, in extreme cases, bring about landslides. Finite resources form over very long periods of time and cannot be replaced in a human lifetime. Therefore, they must be used responsibly and be recycled if possible.

Organic materials

Modern developments in the plastics industry have seen the introduction of bioplastics or biopolymers. Biopolymers are plastics that are made from renewable resources such as starch, sugar and cellulose sourced from animals or plants and are ideal for disposable items such as carrier bags and packaging as they can decompose and the nutrients return to the soil as part of the cradle-to-cradle circular economy.

Renewable resources

Renewable resources are the materials that we have the capability to produce, such as wood or biopolymers. For example, wood is said to be a renewable resource as a new tree can be planted to replace one that is cut down. This is a sustainable approach to managing materials, ensuring wood will be available for years to come. However, if the trees are not replaced, use of wood becomes unsustainable and such harvesting causes irreparable environmental damage.

Case study

Sustainable packaging materials

In addition to considering the sustainability of the materials used for the product, designers must consider the environmental impact of packaging. Pallets, crates and packing trays are commonly reused when transporting products and traditional packaging materials, such as paper bags and cardboard boxes, can be recycled.

Some plastic carrier bags take at least 20 years to decompose. However, the plastic bags now provided by most local councils for food waste are biodegradable. These are made from corn starch and a biopolymer called polylactide (PLA). The bags decompose along with the food waste.

Biopolymers make an extremely sustainable packaging material as they use renewable resources and are not harmful to the environment when they are disposed of correctly. Other biodegradable packaging materials include loose corn starch 'peanuts' that replace polystyrene balls, bubble wrap and air bags. Fungi or mushroom packaging is made from waste cornstalks and seed husks mixed with fungal mycelium, which grows and bonds the corn material. This material replaces styrofoam and polystyrene, and can be used to secure the contents, such as computers, in packaging boxes.

Recycling materials

Assuming that all the materials used in the product are recyclable and that the user has taken the time to separate the materials correctly, then the product's materials can be used again. This is the best case scenario. The worst case scenario is when the materials used in the product are not recyclable (some manufactured boards for example). Or, when the user does not recycle and the product ends up in a landfill.

Landfill waste is buried and compacted to rot away over hundreds of years.

MATERIALS AND MANUFACTURING

> **Make the Link**
>
> Plastic products should be marked with a recycling symbol to identify their type of plastic, making them easier to recycle.

Cradle-to-grave

When products are designed without any consideration of how the materials can be reused, it is said that these are *cradle-to-grave* products. These products use raw materials that either cannot be recycled, are very difficult to recycle or the parts of the product are difficult to separate for recycling. They waste our natural resources and cause pollution in landfills. Such products are part of a **linear economy** because they reach an end point, and this type of manufacturing is sometimes referred to as 'take, make, dump'.

Cradle-to-cradle approach

The term 'cradle to cradle' is used to describe design and manufacture in total harmony with the Earth; whatever we take from the land must eventually be returned to nature. This approach is about designing for disassembly, to allow the separation of the materials, so they can be reused, reprocessed or recycled. The materials can be cascaded through different uses until they can be returned to nature. Therefore, when selecting an appropriate material, the entire life cycle of the product must be considered.

Some materials, such as biopolymers made from corn starch, are designed for the **biological cycle**. These materials are biodegradable and are safely returned to the Earth through composting as they contain no harmful toxins.

When all the materials can be reused or returned to nature, this is called a 'closed loop' or 'circular-economy approach'. This means that products must:

These seedling pots will decompose after the seedlings have grown, making them a sustainable biological material.

- be 100% recyclable
- not damage the Earth with any chemicals or toxic waste
- not disturb or damage the Earth's ecosystem
- be manufactured using renewable energy sources.

The cradle-to-cradle approach can preserve our natural resources, minimise air pollution and have an overall neutral or positive impact on the environment.

AN INTRODUCTION TO MATERIALS

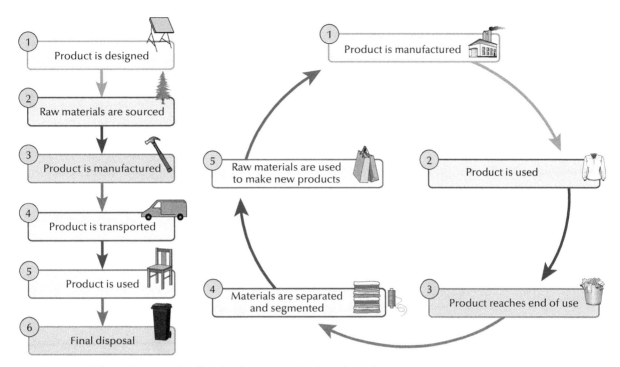

Products can follow a linear path whereby they are not broken down for reuse or recycling or they can follow a circular life cycle in which they are reused or recycled. The circular path is said to be sustainable as it reuses materials instead of wasting them.

Upcycling materials

It is popular for designers to use materials from discarded products instead of using raw materials. This is called **upcycling**, as it gives old products a new lease of life. It is an extremely good way of working sustainably as it incorporates the **six Rs** of recycling:

- **Rethink**
- **Reuse**
- **Recycle**
- **Repair**
- **Reduce**
- **Refuse!**

Upcycling is a positive approach compared to recycling. With recycling, materials can degrade in quality each time they are recycled. Recycling may also involve energy and water consumption, and often transportation.

Case study

Forestry Commission Scotland
Coimisean na Coilltearachd Alba

Scotland's forests are the most productive in the UK. They make a significant contribution to Scotland's economy through jobs in the wood-processing industry, forest management, wood transportation and other associated industries.

Forestry Commission Scotland (FCS) aims to maximise the economic potential of Scotland's timber resources. It encourages continued investment in timber processing by sustaining a predictable and stable supply of good quality timber. It plants 24 million trees every year, to create new woodland and to replace the trees harvested. Some of these trees will help to regenerate blighted industrial landscapes, such as former coalfield communities, and to bring new woodlands closer to urban areas.

A sustainable forest is a forest that is carefully managed, so that as trees are cut down they are replaced with seedlings that eventually grow into bigger, mature trees. The forest provides raw materials for furniture and construction, and wood pulp for paper. Great care is taken to ensure the safety of wildlife and to look after the natural environment.

Sustainable forests make business and environmental sense, as raw materials can continue to grow, they provide habitats for wildlife and they also attract walkers and hikers.

Sustainability is at the heart of FCS, setting the standards for the sustainable management of the UK's forests based on internationally recognised science and best practice. Britain was the first country in the world to have all its public forests independently certified as being sustainably managed.

FCS sustainably harvests almost five million tonnes of wood every year from Britain's public forests. That's around 44% of total domestic production or 300 truckloads every day. This reduces our dependency on imported wood and provides low-carbon materials for domestic wood-using industries, and for fuel and energy. The income from timber helps to offset the costs of managing the forests.

As Britain's largest land manager, FCS is custodian of one million hectares of land, including some of our best-loved and most spectacular landscapes. Two-thirds of FCS estates lie within national parks, areas of outstanding natural beauty or sites of special scientific interest (SSSIs).

Check your progress

I can: HELP NEEDED   GETTING THERE   CONFIDENT

- describe the properties and identifying features of a range of woods and manufactured boards   ◯ ◯ ◯

- describe the properties and identifying features of a range of metals   ◯ ◯ ◯

- describe the properties and identifying features of a range of plastics   ◯ ◯ ◯

- explain why a material is suitable for the manufacture of a component or product   ◯ ◯ ◯

- describe the sustainability issues related to materials.   ◯ ◯ ◯

MATERIALS AND MANUFACTURING

5 Manufacturing in the workshop

> **By the end of this chapter you should be able to:**
> - describe methods of preparing for manufacture
> - state the names of appropriate tools, equipment and manufacturing processes for working with wood
> - state the names of appropriate tools, equipment and manufacturing processes for working with metal
> - state the names of appropriate tools, equipment and manufacturing processes for working with plastic.

> **Make the Link**
> The cutting list (page 92) is an essential part of the plan for manufacture (page 90).

Prepare for manufacture

Once a design proposal has been finalised and a plan for manufacture has been completed (see page 90) it is time to begin manufacture.

Purchasing materials and components

A designer will work with the engineer to create a cutting list to order materials for the manufacturer. Without a cutting list, no materials can be ordered.

> **Make the Link**
> Quality assurance checks are completed by the manufacturers in industry.

Quality assurance

When you receive your materials, you must check them over to assure their quality. This involves checking for any flaws in the material and assessing its general overall quality. This is the best time to identify any faults with the material as it is easier to fix or replace at this stage. If you have ordered components then they should be checked for dents, manufacturing flaws and that the correct amount have arrived.

Standard components

Standard components are the common parts that are used in products. To design every part from scratch would overcomplicate the design, waste time and money. Standard components make the manufacture of a product easier, and also make it easier to get replacement parts. Standard components include wheels, brackets, fitting and fixing components, washers, handles, electronic parts, etc.

MANUFACTURING IN THE WORKSHOP

Standard components, also called 'knock-down fittings' are commonly used to join flat-pack furniture. This keeps the cost of the product low as the furniture is not built by a skilled worker in a factory. Also, the size of a factory can be reduced if assembly space isn't required and the cost of rent, electricity and power is therefore reduced. It is quicker for the furniture to be packaged up and sent to the shop and more boxes can fit into a lorry. Many people enjoy building flat-pack furniture as there is a feeling of satisfaction from building a piece of furniture and seeing the final result. It can be irritating to the consumer when there are parts missing however, or if the person building the furniture lacks skill or the instructions are inadequate.

This design includes wheels as a standard component.

Knock-down fittings are used to build flat-packed products.

> ### ✔ Test your knowledge
>
> **Standard components**
> 1. Explain the term 'standard components'.
> 2. Give an example of a product that has used standard components.
> 3. State three advantages of knock-down fittings.
> 4. State three drawbacks of knock-down fittings.

Sequence of operations

The sequence of operations, also known as the plan for manufacture, is used to guide you through the construction of the product. It will include:

- steps and order of tasks
- tools and machines being used
- safety references.

Steps and order of tasks

The order of tasks can be difficult to predict, as the project is new to you and you may be unfamiliar with some of the processes involved. Sometimes the sequence of manufacture is simple. For example, the hole must be drilled in a clock before the clock mechanism can be assembled. Sometimes the sequence of manufacture is more complex. For example, in this key cabinet (below) the plastic backing had to be cut on a laser cutter before the project was assembled. This required further thought as the key cabinet was to be painted and, therefore, the plastic had to be carefully covered with masking tape before and during painting.

Key cabinet before assembly and then masked off before painting.

It may be essential to research the sequence of a process. Some processes, such as dip coating metal with the Fluidiser and smoothing the edges of plastic, have a very specific sequence of manufacture.

Once a sequence for manufacture has been decided, you should be able to justify the reasons for the order of tasks. The sequence must be the most practical method, in that it must be easy to put into practice, and it must be the most efficient method, in that it must make good use of time and resources.

Some reasons to justify your sequence may be:

- The process was researched and this sequence was found to be a common approach.
- A practice or test was carried out to investigate the best possible method.

MANUFACTURING IN THE WORKSHOP

- Common sense was applied.
- Time can be used more effectively by following this sequence.
- Grouping similar tasks together saves time.
- The time taken to complete this sequence is the quickest.
- The quality of work from this sequence will be the best.

The working drawings are a useful resource in the workshop. They can be used to find the sizes when marking out and they can be used to select tools and equipment for manufacturing tasks.

Tools and machines being used

For successful manufacture, it is important to learn and use the correct names for tools, equipment and processes. This will help you to communicate effectively the manufacturing processes and techniques you will use during manufacture.

When preparing for manufacturing a large wooden product, it may be necessary to build up wide panels of wood by gluing planks of wood together with PVA glue. When writing the sequence of operations, it would look like this:

For example, something like this provides just enough information:

Step	Process	Tools
1	Glue together planks with sash cramps	Sash cramps

A more detailed version would look like this:

Step	Process	Tools/Resources	Notes	Timing
1	Glue together planks to make wider boards. Lay three sash cramps down on the bench, arrange wood on the cramps. Apply glue to edges. Use scrap wood to protect the edges from being dented.	Sash cramps PVA glue Lengths of scrap wood Paper towels		1 hour

MATERIALS AND MANUFACTURING

> **Hint**
> It can be useful to include the time you plan to spend on each task, particularly for your Course Assessment Task, which includes a timed manufacturing project.

Planks of red pine are glued together to create a wider board. The sash clamp on top prevents the boards from bowing while the scrap wood at each end protects the red pine from being damaged when the clamps are tightened.

Safety references

Before starting any practical work, it is extremely important to understand the safety implications of a project. For example, when using an oven to shape a sheet of plastic, gloves are worn to protect your hands from the heat.

Special safety equipment must be worn in many different jobs – people from lumberjacks to doctors wear PPE. It is a responsible approach to avoid accidents at work and employers will insist PPE is worn for the health and safety of their workers. PPE is worn in the workshop too:

- Safety goggles protect eyes when using machinery.
- A face mask is used when sanding wood or when using toxic finishes such as spray paint.
- A face shield is required when working with hot metal.
- Heatproof suede apron and gloves are needed when working with hot metal and plastic.

Special safety equipment must be work in many different jobs from lumberjacks to doctors. It is a responsible approach to avoid accidents at work and employers will insist it is worn for the health and safety of their workers.

The general safety rules for a workshop are:

- Keep jackets and bags out of the work area, to avoid tripping hazards.
- Tuck in all loose clothing, earphones and jewellery and tie back long hair.
- Keep tools in the middle of the bench or put them away when they are not in use.
- Report any breakage or accident, even if very small.
- Pay proper attention during demonstrations to ensure you understand how to safely use the tool, machine or equipment.
- Wear safety goggles when using machines.
- Observe the safety zones marked out on the floor for machines.
- Use guards on machines.
- Set up machines with the isolator switch (main power) off.
- Keep hands away from moving parts of machines.
- Ensure drill bits and other machine parts are secure before using machines.

Examine tools and equipment before use to check they are in safe working order. Using faulty or broken tools and equipment is very dangerous and can lead to accidents. There are some faults and breakages with tools and equipment that are common:

- File handles can come loose or come apart completely.
- Plane blades are adjustable and so can often become loose or angled.
- Mallet heads can come loose from the handle or come apart completely.
- Coping saw and jigsaw blades can snap.

Make the Link

Safety in the workshop – later on in this chapter.

Something like this provides just enough information:

Step	Process	Tools/Resources
1	Draw round the template	Template and pencil
2	Cut sides to shape	Jigsaw
		Dust mask
3	Smooth with sandpaper	Sandpaper
		Dust mask

MATERIALS AND MANUFACTURING

A more detailed version would look like this:

Step	Process	Tools/Resources	Notes
1	Draw round the template	Template and pencil	
2	Secure to the bench and cut to shape using a jigsaw. Attach the vacuum cleaner hose to the bench to aid dust extraction.	Jigsaw Dust mask G clamp Vacuum cleaner	G clamp the vacuum cleaner to the bench during cutting
3	Smooth with sandpaper	Sandpaper Dust mask	Keep vacuum cleaner on for prolonged use

The signs you see around the workshop are colour coded: blue signs are for mandatory rules (must do), red signs identify a danger, yellow signs are to warn you of a hazard and green signs are for safety and first aid.

Activity

Sequence of operations

A pupil has designed a letter rack like the one shown here.

The steps for the manufacture of the letter rack were written down in the wrong order:

- Mark out the shape on the sheet metal.
- Round the edges of the metal sheet.
- Smooth any rough edges or dribbles of primer with wet-and-dry paper.
- Apply the first coat of paint.
- Apply the second coat of paint.
- Draw file the edges of the metal sheet to create a smooth edge.
- Mark out the holes.
- Form the shape of the metal by bending it.
- Use emery paper to finish the edges of the metal.
- Cross file the edges of the metal sheet to remove any bumps.
- Apply the primer using a brush.
- Cut the sheet metal to the hand shape using a junior hacksaw
- Smooth any rough edges or dribbles of paint with wet-and-dry paper.

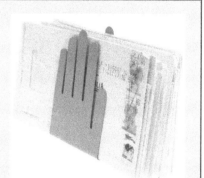

1. Group the tasks into the following **four** headings:
   a) Marking out
   b) Cutting, shaping and drilling
   c) Smoothing the edges of the metal
   d) Finishing
2. Reorganise all the steps into the correct sequence.

Manufacturing with wood

When working with wood and manufactured boards, the main processes are:

- **Cutting, sizing and shaping**
- **Drilling**
- **Turning**
- **Assembly and joining methods**
- **Finishing**

Cutting, sizing and shaping wood

Cutting, sizing and shaping wood requires a range of different tools, depending on the shape and size of the project. First, the raw material may need to be cut to size using a saw, simply called sizing. Wood can then be measured and **marked out**. This involves accurately drawing the component parts or woodwork joints onto the wood or manufactured board. The table below shows some of the most common tools for marking out.

Tool	Image	Use
Tape measure		Checking material sizes
Steel rule		Measuring
Try square		Drawing lines at 90° to the edge of the wood or checking for 90° angles
Sliding bevel		Similar to a try square, but adjustable to any angle

Continued

MATERIALS AND MANUFACTURING

Tool	Image	Use
Marking gauge		Drawing lines parallel to the edge of the wood
Mortise gauge		Similar to a marking gauge, but with two pins for marking mortise joints
Bradawl		Marking hole positions

> **Hint**
> A coping saw can cope with curves.

Once it is marked out, the material can be cut using saws that are suitable for wood. A tenon saw or mitre saw is used for straight cuts, while a coping saw can cut curves. The table below shows some of the most common tools used for cutting wood.

Tool	Image	Use
Tenon saw		Sawing straight cuts
Coping saw		Sawing curved cuts
Mitre saw		Sawing at a specific angle, with a range of angles

To cut woodwork joints, a chisel can be used with a mallet. There are many variations of plane, which can remove waste material, smooth surfaces or cut grooves. These are some of the most common tools used for shaping wood.

MANUFACTURING IN THE WORKSHOP

Tool	Image	Use
Jack plane		Adjusts the thickness of wood by removing a layer of material
Smoothing plane		Smooths wood by removing a thin layer of material
Rebate plane		Removes a thin section of wood (rebate) along one edge
Mortise chisel		Cutting out wood, especially suited to mortise joints
Bevelled-edge chisel		Cutting out wood
Mallet		Used to hit a chisel

✓ Test your knowledge

Woodwork tools

1. State three tools used for measuring wood.
2. Explain the difference between a marking gauge and a mortise gauge.
3. Name the saw that can be used to cut a curved edge on a manufactured board.
4. Explain one reason for using a jack plane
5. Name a tool suitable for levelling a housing joint.

Machines and power tools for wood

A range of machines and power tools can be used to shape wood. These make manual tasks faster and easier. However, extra safety precautions must be followed when using them:

- Wear appropriate PPE (e.g. wear safety goggles when using the mortise machine).
- Test the emergency stop button before using a machine.
- Keep a safe distance from others.
- Be aware of the power cable when using power tools.
- Keep fingers away from sharp or moving parts.

Card templates are a quick way of marking out complex or curvy shapes and they can be reused, allowing for small batches of identical products to be produced.

MATERIALS AND MANUFACTURING

Each workshop has different machinery. These are some of the most common workshop machines.

Tool	Image	Use
Woodwork lathe		Creates cylindrical wooden shapes by turning wood
Sanding machine		Smooths the edges of wood
Band saw		Large machine saw for cutting materials
Fret/Scroll saw		Smaller saw used for thin material, and intricate cuts and tight curves
Mortise machine		Cuts square or rectangular holes or slots in wood

Hint

The pillar drill can be used for drilling wood, metal and plastic.

Power tools can be used to cut and shape material when greater flexibility is required, where a machine is unsuitable or to speed up a manual task. When working with power tools, always take care where you place your hands and ensure that the power cable is held out of the way. The table below shows some of the most common power tools.

MANUFACTURING IN THE WORKSHOP

Tool	Image	Use
Orbital sander		Sands wood to an extremely smooth finish
Power drill		General drilling, can act as a screwdriver with a screwdriver attachment
Jigsaw		Cutting curved cuts or quick rough cuts
Biscuit jointer		Cuts a groove for a biscuit joint
Router		Cuts a curved profile along an edge; can also cut a rebate

✔ Test your knowledge

Machines and power tools for wood

1. State five safety rules when using power tools and machines.
2. Describe the purpose of the sanding machine.
3. Name a machine that can cut a square slot into wood.
4. Explain one benefit of using an orbital sander.
5. Outline two safety checks when using a jigsaw.
6. Name an alternative use for the power drill other than drilling holes.

Drilling wood

To make a round hole in wood, a drill is used. A hand drill, brace, power drill or pillar drill can be used to drill wood. There is also a range of different drill bits suitable for drilling wood, depending on the shape and size of the project. For example, a forstner bit will drill a flat-bottomed hole. A steel rule and try

MATERIALS AND MANUFACTURING

square or marking gauge should be used to mark out the location of the hole accurately. The table below shows some of the most common tools used for drilling wood.

Tool	Image	Use
Twist drill bit		Drilling general holes
Forstner drill bit		Drilling flat-bottomed holes
Countersunk bit		Drilling countersunk holes
Hole saw		Drilling and creating round discs

When marking out a hole, a cross is marked on the material, allowing for the tip of the drill bit to line up with this exact point.

The pillar drill (below) is ideal for most woodwork drilling tasks and can also be used to drill metal and plastic. While using this machine, the material must be clamped to the table using a G-clamp. A machine vice can be used to hold small pieces of wood. However, the vice can dent the wood if it is overtightened. The drill should be adjusted to suit the depth you want to drill to using the depth stop and/or adjusting the table height. A small hole is drilled first. This is called a **pilot hole** as it guides the way for a larger hole to be drilled and prevents the material from cracking. A pilot hole is usually 1–2mm in diameter.

The same process is followed for drilling plastic.

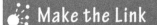 Make the Link

The pillar drill is also used for drilling metal and plastic.

🔍 Hint

Use a small piece of scrap wood under the G-clamp to avoid denting wood. Remember to remove the chuck key as this can cause a serious accident if left in the machine.

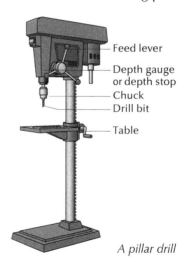

A pillar drill

MANUFACTURING IN THE WORKSHOP

✔ Test your knowledge

The pillar drill

1. Name the drill bit required to drill a flat bottomed hole in wood.
2. State three safety rules when using the pillar drill.
3. Explain the purpose of the depth gauge.
4. Describe the purpose of a pilot hole.
5. Name a tool that can be used to secure the wood to the table while drilling.
6. Name a drill bit suitable for a pilot hole.
7. Explain the process of drilling an 8-mm hole in a piece of wood.

Turning wood

The wood turning lathe is used for turning between centres (to make long cylindrical shapes such as table legs), or for face plate turning (using round blocks of wood to make e.g. bowls). To prepare for using a lathe, a length of wood, known as a blank, is marked out and planed to a cylindrical shape.

Make the Link

When selecting wood for turning, look for a piece of wood that is free from knots.

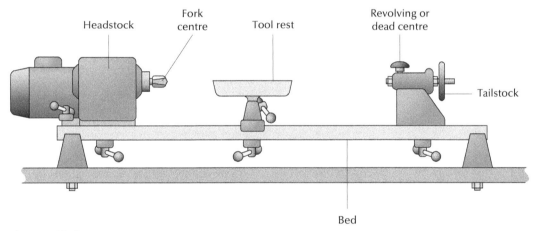

The wood lathe

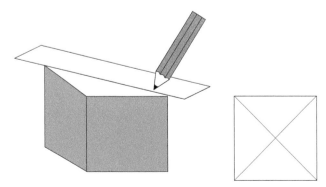

Step 1: Draw corner to corner on each end of the blank using a pencil and steel ruler.

MATERIALS AND MANUFACTURING

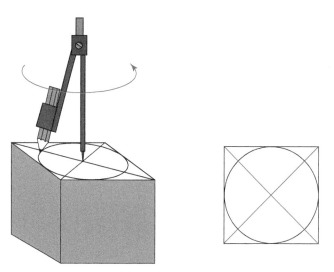

Step 2: Use a compass to draw a circle on each end of the blank. The circle should be big enough to touch the edges of the blank.

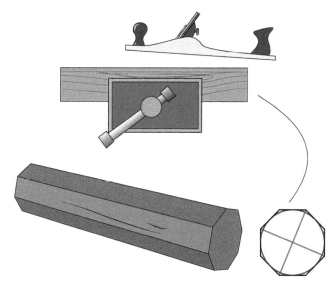

Step 3: Plane the blank to an octagonal shape or as round as you can without planing into the circle at each end.

Step 4: Use a tenon saw to cut a groove on one end of the blank.

It is also possible to use the woodworking lathe for face plate turning (round blocks of wood), such as bowls.

When wood is manufactured on a wood turning lathe, the process is called turning as the machine spins the wood around. The wood is held tightly between the fork centre and the revolving or dead centre. Before using a wood lathe, the following checks must be made:

- The material is held tightly on the machine.
- No loose clothing or jewellery is in the way.
- The speed is set correctly for the task.
- The material can turn freely without hitting any machine parts.
- Safety goggles are worn.
- The material is held exactly in the centre.
- The tailstock is secured or locked in position.

Once prepared, the blank is then held between the fork centre on the headstock and the dead centre or revolving centre at the tailstock end. The operator holds a tool on the tool rest and can shape the wood by moving it in different directions. A range of curves, indents, angles and shapes can be created with a range of different wood-lathe tools. First, it roughly rounds the product off, then other shaping tools can be used to create more detailed shapes. The key aspects of using the wood lathe are:

- Shaping – using lathe tools to create curves, rounded features and beads.
- Parallel turning – running the lathe tool along the wood (parallel to the wood) to create cylinders.

Make the Link

Chapter 6 explains wood turning in mass manufacture.

- Parting off – for cutting a narrow slot and reducing the diameter of the wood at each end to make it easier to remove waste material.
- Finishing – removing the tool rest and holding sandpaper on the wood to smooth it while the machine is on.

The table below shows some of the most common tools used for turning.

Tool	Image	Use
Skew chisel		Sharp-edged tool for accurate rounding and shaping
Parting tool		Pointed tool for cutting notches or cutting/removing wood from the lathe
Rounded scraper		Rounded-edge tool for removing wood quickly and creating convex curves
Gouge		U-shaped tool for removing waste wood quickly and creating convex curves
Outside callipers		Measuring outside diameters on the woodwork lathe

✔ Test your knowledge

Wood turning

1. State the checks that are made on the machine before using a wood turning lathe.
2. State the checks that are made on the operator before using a wood turning lathe.
3. Describe the **four** stages required to prepare the wooden blank before turning.
4. State the name of a suitable tool for rounding and shaping wood on the wood turning lathe.
5. State the name of a suitable tool for cutting notches on the wood turning lathe.

Link to suggested answers
www.collins.co.uk/pages/scottish-curriculum-free-resources

Assembly and joining methods for wood

There is a range of woodwork joints. Designers select woodwork joints based on their:

- strength
- aesthetic qualities
- functional suitability
- ease of manufacturing.

Most woodwork joints are held together by an adhesive to strengthen the joint. However, panel pins or screws can also be used to reinforce a joint. This **rub joint** has a small surface area with just two surfaces to coat with adhesive, therefore the joint will be weak. Rub joints such as **mitre joints** and **butt joints** are simply two surfaces glued together with no reinforcement. A **lap joint** has more surfaces that interlock, therefore it has a larger surface area to coat in adhesive, therefore the join will be stronger. Aesthetic qualities of a joint can be that they are simple or that the joining method is hidden. A **stopped housing joint** hides the joint, simplifying the overall look and improving the aesthetic of the product. Aesthetic qualities of a joint can be that they are simple or that the joining method is hidden.

Some woodwork joints are more complex to manufacture. However, they can be more attractive as their complexity and detail indicates craftsmanship and skill.

The most common wood adhesive is PVA, which is a white glue that dries clear. It is a non-toxic, non-flammable, water-based adhesive. As it is waterproof, any excess glue must be wiped away with a damp cloth or paper towel before it dries. It is, however, unsuitable for outdoor use. It is best to leave PVA for around 24 hours to dry properly.

Adhesives that are suitable for outdoor use are harder and more durable. However, they are generally toxic and must be used with care.

Joints range in terms of their strength and difficulty of manufacture. The correct joint to choose will depend on the function of the frame that is being constructed; the wooden framework for a building obviously has different requirements to that of a picture frame. The table on the next page shows some of the most common framework joints.

Cabinet joints also have a range of strength and difficulty. Cabinet joints can sometimes be called **carcass joints**. The table that follows shows some of the most common cabinet joints.

> **Make the Link**
>
> Wood and commercial manufacture: Chapter 6 explains wood turning in mass manufacture.

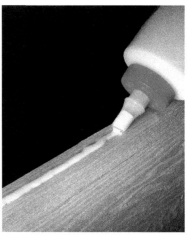

The most common wood adhesive is PVA.

Joint	Image	Strength	Manufacture
Butt		Very weak	Easy and quick
Mitre		Weak	Easy and quick
Lap		Moderate	Fairly easy and quick
Dowel		Moderate	Moderate difficulty and time requirement
Halving		Moderate	Moderate difficulty and time requirement
Bridle		Strong	Challenging and time consuming
Mortise and tenon		Strong	Challenging and time consuming

When joining materials, they should be **dry clamped** first. This involves putting the whole project together without glue to check the assembly.

Joint	Image	Strength	Manufacture
Butt	See table opposite	Very weak	Easy and quick
Dowel	See table opposite	Moderate	Moderate difficulty and time
Through housing		Moderate	Moderate difficulty and time requirement
Stopped housing		Strong	Somewhat challenging and time consuming
Finger		Extremely strong	Challenging and time consuming
Dovetail		Extremely strong	Very challenging and time consuming
Biscuit		Moderate	Easy and quick but requires machinery

When assembling, the four aspects to check for a good quality assembly are that the project is:

- **Square** – check the corners with a try square or measure the diagonals and check they are equal.
- **Level** – check the project is sitting flat on the surface and is not wobbling. You can also check horizontal parts are level using a spirit level.
- **True** – check all the parts are correctly and accurately aligned.
- **Secure** – check that all parts are properly fitted together, especially wood joints and boards that fit into slots. Use extra clamps when necessary for extra accuracy.

In the table on the next page are some common tools that are used in the assembly of products.

> **Hint**
> Before gluing any wood, clamp it together without glue to check it is square and fits properly. This is called dry clamping.

When marking out a joint, use cross hatching to show the waste wood.

> **Hint**
> Put your project on a large flat board to avoid assembling on uneven workbenches.

MATERIALS AND MANUFACTURING

Type of hammer	Image	Usage
Claw hammer		Nailing or removing nails or panel pins; also good for levering items apart
Cross pein hammer		Nailing small nails or panel pins; the smaller end is used first when holding the nail/pin, to avoid hitting your hand
Screwdriver		Attaching screws
Nail punch		Drives nails below the surface of the wood for a better look and tighter join
G-clamp		Holds wood in place while cutting or for assembly
Sash cramp		Holds wood in place during assembly

> **Make the Link**
> Knock-down fittings are explained earlier in this chapter.

A range of fittings and fixings is available for woodworking projects. Here are some of the more common items.

Type of fitting/fixing	Image	Usage
Roundhead screws		Rounded tops have aesthetic qualities
Countersunk screws		Sit flush (level) with the surface, hiding the screw
Panel pin		A general purpose nail
Lost head nail		Nails that can be hit below the surface of the wood
Butt hinge		Classic hinge that must be let-in (set in a chiselled-out recess)
Piano hinge		A long-length hinge
Knock-down fittings		A range of joining fittings used as easy alternatives to joints

MANUFACTURING IN THE WORKSHOP

✓ Test your knowledge

Assembling and joining methods

1. Name two rub joints.
2. State the type of glue used to join wood.
3. Name three moderately strong frame joints.
4. Explain the difference between a stopped housing and a through housing joint.
5. Describe how waste wood should be marked onto wood.
6. Explain the term 'dry clamping'.
7. Name the four aspects to check for a good quality assembly.
8. Name a tool that can be used to hit panel pins into a backboard.
9. State two benefits of using a nail punch to drive nails below the surface of the wood.
10. Explain the difference between roundhead screws and countersunk screws.
11. State three benefits of using a knock-down fitting instead of traditional woodwork joints.

Finishing wood

When deciding on the type of finish for a product or component part, it is important to consider the following factors:

- **aesthetic**
- **function**
- **lifespan** required
- **location** (interior or exterior)

Certain wood finishes will make the wood more durable or waterproof, while others will simply offer a different colour or shine.

Well-prepared surfaces create a good quality finish. Sandpaper, also known as glasspaper, is a type of abrasive paper used to remove small amounts of material from surfaces, to make them smoother, remove pencil marks or remove dried glue. Sandpaper is available in different grades from coarse, which feels rough to the touch, to fine, which is smoother.

Always sand *with the grain* of the wood (in the direction of the lines).

Sanding across the grain will scratch your wood and damage it. When sanding the end grain (where the annual rings are), sanding in a circular motion will achieve the best finish.

Belt and disc sanding machines can smooth the surfaces and edges of wood to a very high standard. The sandpaper moves extremely quickly and it provides a much faster finish than hand sanding. However, a final sand by hand or using an orbital sander gives a higher quality finish.

Here are some of the most common pieces of equipment for applying a finish.

> 🔍 **Hint**
>
> Check the edges of the wood for diagonal saw marks (left by the circular saw) and remove them.

Sand the wood until it has the same level of smoothness all over using a sanding block, or cork block.

147

MATERIALS AND MANUFACTURING

Equipment type	Image	Usage
Brushes		Applying paint, varnish, stain or sanding sealer
Steel wool		Smoothing between varnish coats or for applying wax
Sanding block		Sandpaper is wrapped around it, providing even pressure
Sand/glass paper		Abrasive paper for smoothing wood
Cloth		Applying wax, oil or polish

A natural finish can be achieved by simply sanding the wood smooth.

Using wood stain is a clever way to make a low-cost wood appear more expensive or to add colour.

 Hint

Avoid applying half a coat as it can appear patchy when it dries.

Oil will enhance a natural finish or protect the wood. There is a range of different types. Vegetable oil is safe to use on products that will be in contact with food. Danish oil and teak oil add a depth of colour and contain minerals that help to preserve wood. To prepare for oiling, the wood must be clean and clear of dust. Oil is applied with a clean, dry cloth or with steel wool, using quick strokes up and down the grain. Oiled wood must be left for about 24 hours to properly dry but will only need one coat, if done properly.

Wood stain requires extra preparation as it is water-based. This extra preparation is called *raising the grain*. Before applying the stain, the wood is dampened with a wet cloth. This makes the wood swell up. Once dry, the wood will be slightly plumper and so it is sanded back down to create a level and smooth surface. The stain can then be applied with a clean, dry cloth using quick circular motions or with a brush. Most stained wood can be lacquered or waxed to give a glossy finish and protect the wood; however, be aware that some stains react badly with lacquer or wax.

Varnish is applied with a brush. Dip the brush into the finish and wipe off the excess on the side of the tin. Brush a thin layer in the direction of the grain and allow plenty of time to complete one full coat. Leave lots of time between coats for the finish to dry properly. Before applying a second coat, give the wood a very light sand with fine sandpaper to remove any bumps or flaws. Use sandpaper to remove any bristles that have dried into the finish. Steel wool or fine grade sandpaper is excellent for smoothing varnish between coats.

Paint is also applied with a brush, using the same method as varnish. Paint can also be applied with a roller. Use a roller tray to lightly coat the roller with paint. Cover a thin layer in one direction and leave to dry. The paint roller can be wrapped tightly in a plastic bag and used again the next day. Again, use very light sand with fine sandpaper to remove any runs, bumps or flaws before applying a second and third coat.

Wax is a common finish used to bring a high quality finish and extra shine to wooden products. There is a range of colours and types of wax available. These can be used to enhance the shine or to protect untreated wood. To apply wax, a base layer of sanding sealer (a watery sealant) or a light varnish is advised to provide a base for the wax. Once the wood is dry, ensure it is clean, then apply the wax with a clean, dry cloth or steel wool, using quick circular motions to give an even finish.

Lacquer is a glossy, clear finish. Traditionally, lacquer was applied with a brush. However, this can leave brush marks on the surface and spraying lacquer onto wood achieves a smooth and glossy finish.

Many mass manufactured wooden products have a lacquered finish as it is the fastest drying finish.

Before using a spray lacquer, open all the windows in the room for ventilation as the fumes are toxic and dangerous to inhale. To apply lacquer from an aerosol can, first read the directions on the can. Usually, the can must be about 300 to 400mm away from the wood. Spray in even strokes, coating thinly.

As the finish dries quickly, this can be repeated almost immediately, coating the surface with many more thin layers. To apply lacquer with a brush, follow the procedure for painting or varnishing.

✔ Test your knowledge

Link to suggested answers
www.collins.co.uk/pages/scottish-curriculum-free-resources

Finishing wood

1. State three benefits of applying a finish to wood.
2. Name five wood finishes.
3. Explain why vegetable oil is a suitable finish for a wooden fruit bowl.
4. Describe the process of 'raising the grain' when applying a wood stain.
5. State four stages of applying varnish to a woodwork project.
6. Name a method of applying paint without using a brush.
7. Describe how to apply lacquer to wood.

MATERIALS AND MANUFACTURING

Manufacturing with plastic

Plastic has a bright, colourful and shiny appearance, which makes it appealing to use. However, it can be a tricky material to work with. It cracks and snaps under pressure during manufacture and it can scratch easily. Most plastic sheets come with a plastic covering to protect them and this should be kept on for as long as possible.

When working with plastic, the main processes are:

- **Cutting and shaping**
- **Drilling**
- **Forming, bending and twisting**
- **Moulding**
- **Finishing**

Cutting and shaping plastic

> **Hint**
> Keep the protective film on sheets of plastic for as long as possible, to protect the plastic from scratches.

When cutting and shaping plastic, care must be taken as it can easily crack or snap. However, there are many techniques that can be used to avoid damaging the plastic during manufacture.

To mark out on plastic, a non-permanent marker or chinagraph pencil is required; a pencil will not work. Tools that scrape into the plastic, such as a scriber or a marking gauge, can be used to mark out, but only for waste plastic. A compass can be used for drawing circles or curves. However, the point will also scratch the plastic, so it is common to use a template for this job. The most common tools for measuring and marking out on plastic are shown here.

Tool	Image	Usage
Tape measure		Checking material sizes
Steel rule		Measuring
Scriber		Drawing lines on waste material

Plastic curves are cut using a coping saw, while straight cuts are cut with a junior hacksaw. These saws leave a 'raw edge' on plastic, and so files are used to smooth away the rough edges. Then the edges can be finished using wet-and-dry paper and a cloth with polish on it. The most common tools and workshop equipment used in plastics work are shown in the table opposite.

A pupil has covered their plastic with masking tape to mark out a line parallel to the edge of the plastic using a marking gauge.

Tool	Image	Usage
Coping saw		Sawing curved cuts
Junior hacksaw		Sawing at a specific angle, with a range of angles
File		Smoothing the edge of plastic
Wet-and-dry paper		Abrasive paper for smoothing plastic
Cloth		Applying polish

The edges of plastic are smoothed using the four stages (CDEF):

- **C**ross file to remove any large bumps, by moving the file across the material.
- **D**raw file for a smooth finish after cross filing, by drawing the file along the material.
- **E**ven out with a wet piece of wet-and-dry paper to make it super smooth and shiny.
- **F**inish with a polish.

> **Hint**
> Cross filing and draw filing are *techniques*, not types of files.

> **Make the Link**
> Metal is also filed using these four stages. However, wood should not be filed, as sandpaper and planes achieve a better quality finish.

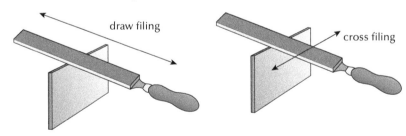

Files are available in a range of shapes and sizes and can be used to shape the edges of the material. For example, a half-round file can be used to file a curve.

Flat   Square   Triangular   Round   Half-round   Knife

A fret/scroll saw or a bandsaw can also be used to cut plastic. However, they leave rough edges. A laser-cutter machine can cut plastic to a very high quality, is very quick and can produce identical products, but these machines are expensive and not always available. The table on the next page shows some of the most common machines used for cutting plastic.

When filing plastic sheets, keep the sheet as low to the bench vice as possible to support the sheet and avoid cracking it, or sandwich the plastic between two pieces of scrap wood in the vice to support the plastic.

MATERIALS AND MANUFACTURING

Machine	Image	Usage
Bandsaw		Large machine saw for general cutting of materials
Fret saw		Smaller saw used for thin material, intricate cuts and tight curves
Laser cutter		Cuts and engraves plastic, wood, paper, glass, metal and fabric sheets

✔ Test your knowledge

Cutting and shaping plastic

1. State the reason why a scriber is only suitable for marking out waste plastic.
2. Name a saw for creating curved cuts on plastic.
3. Name a saw for creating straight cuts on plastic.
4. State the reason for clamping plastic low in a bench vice during manufacture.
5. Name the four stages of smoothing the edges of plastic.
6. Describe the purpose of a half-round file.
7. Outline three benefits of laser cutting plastic compared to traditional cutting methods.
8. State four safety rules to follow when using a fret saw.

Make the Link

The safety recommendations at the start of this chapter outline the safety rules that should be followed when using machines in the workshop.

Joining plastic

Although joints are not used on plastic, it is possible to cut slots in plastic to slot pieces together, like the jewellery tree on the next page. To permanently join plastic to plastic, use an adhesive such as liquid solvent cement, epoxy resin or impact adhesive (superglue). These adhesives are suitable as they can create a strong bond on the relatively non-porous surface found on plastic.

MANUFACTURING IN THE WORKSHOP

To join plastic to wood, a groove can be cut into wood using a multi plane or rebate plane. The blade size can be changed to suit the thickness of the plastic. This technique is used for attaching plastic to picture frames and door panels.

Drilling plastic

To mark out a hole on plastic, place a small tab of masking tape on the plastic and mark it with a pen. During drilling, the drill bit can become hot and the masking tape will prevent any small bits of waste plastic (produced when drilling out the hole) from welding onto the surface of the plastic, giving a cleaner edge.

When drilling plastic, a pillar drill or power drill can be used. However, plastic can easily crack during drilling and so the following precautions must be taken:

- Drill a pilot hole before drilling a large diameter hole.
- Support the plastic with scrap wood underneath it.
- Drill slowly.

> **Make the Link**
>
> Drilling plastic is similar to drilling wood or metal. However, a centre punch should not be used on plastic as this can crack it.

Forming plastic

Thermoplastic becomes soft when heated and then it can be easily bent, formed or twisted into shape. When the plastic cools to room temperature, it sets into the formed shape. With a thermoplastic, if the shape produced is not quite as required, heat can be reapplied and the process repeated. Two ways to heat plastic are to use a strip heater or an oven.

The strip heater

The strip heater can only apply heat along a narrow strip. This allows one bend or fold to be made at a time. These are the four main stages of using the strip heater.

> **Hint**
>
> Remove the protective film from the plastic before using the strip heater.

MATERIALS AND MANUFACTURING

> **Hint**
> Running the plastic under a cold tap will speed up the cooling process.

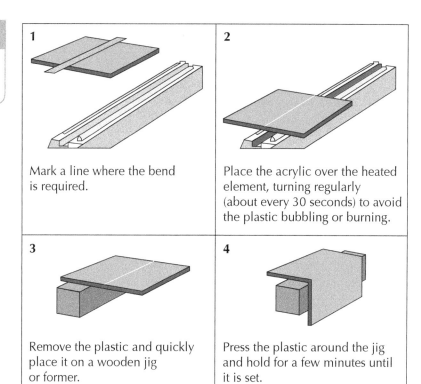

1. Mark a line where the bend is required.
2. Place the acrylic over the heated element, turning regularly (about every 30 seconds) to avoid the plastic bubbling or burning.
3. Remove the plastic and quickly place it on a wooden jig or former.
4. Press the plastic around the jig and hold for a few minutes until it is set.

Wooden formers are blocks of wood manufactured to a particular shape for the plastic to be folded over when it is hot. The former shown above can create a 90° bend, however more complicated formers or different angles can also be created. Without using a former, the fold has nothing to mould itself around and so it is generally less accurate when a former is not used.

Identifying features:

- Plastic bent along a line
- Thin sheets of plastic (around 4mm thick)

The oven

Plastic twists can be achieved by using the oven to soften the plastic.

An oven can be used for more complex shapes. The thickness and type of plastic will determine the time it takes for the plastic to become flexible enough to shape. For example: to shape a 3-mm thick sheet of acrylic, set the oven at a maximum temperature of 170°C and place the plastic in the oven for about 3 minutes, before forming it to the required shape. Heatproof gloves are essential when using the oven. A jig or former can be used to shape plastic into complex shapes, or to make a batch of identical items.

These are the steps to follow when using an oven:

1. Place the plastic in the oven.
2. Leave to heat for a few minutes. Check if the plastic is flexible, close the door quickly if it is not.
3. Once it is flexible, take it out and wrap it round the former.
4. Hold in place to cool.

A former can again be used to shape plastic. With an oven, a more complex shape can be made.

Identifying features:

- Complex 3D shapes
- Sheet, rod, bar or tube of plastic

> ### Hint
> It is easier to finish the edges of plastic before it is formed using a strip heater or oven, as complex shapes can be difficult to hold in the vice.

> ### Make the Link
> Chapter 6 covers the plastic processes that are suitable for mass manufacture.

✔ Test your knowledge

Forming plastic

1. State two ways to heat plastic before it is formed.
2. State the four stages of bending plastic using a strip heater.
3. Name a suitable type of plastic for using on the strip heater.
4. Describe why a wooden former can help to create more accurate forms.
5. State two identifying features of a product made using the strip heater.
6. Outline the stages of using an oven to shape plastic around a former.
7. Name a safety precaution when using the oven.
8. State two identifying features of a product made using the oven.
9. Explain why thermoset plastics are not suitable for forming plastic with the strip heater or oven.

Finishing plastic

Plastics require very little finish as they are waterproof, colourful and are very durable. Polish is the only finish that may be required. This can be applied using a clean, soft cloth.

MATERIALS AND MANUFACTURING

> ### Make the Link
> Mass manufactured plastics can have intricate details (such as symbols, patterns and text) imprinted on products during the moulding process.

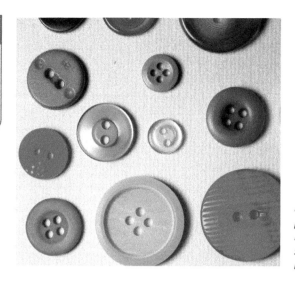

The patterns on these injection-moulded plastic buttons were imprinted on the surface during the moulding process.

Test your knowledge

> **Link to suggested answers**
> www.collins.co.uk/pages/scottish-curriculum-free-resources

Manufacturing with plastic

A pupil has manufactured a set of fridge magnets using workshop tools.

1. A card template was used to mark out the fridge magnets. State **one** benefit of using a card template.
2. Name a suitable saw to cut out the fridge magnets.
3. State the name of a suitable adhesive to glue a small magnet to the back of each piece of plastic.
4. Explain one method to prevent scratches on the surface of the plastic.
5. Describe a method of applying polish to the edges of the plastic.

Manufacturing with metal

Metalwork requires a lot of care and attention as it has many hazards. It is, therefore, important to pay close attention to instructions or demonstrations to learn how to use tools safely.

When working with metal the main processes are:

- **Cutting and shaping**
- **Drilling**
- **Turning**
- **Heat treatments**
- **Casting**
- **Assembly and joining methods**
- **Finishing**

Cutting and shaping metal

There are several ways to shape metal. First, the metal must be measured and marked out. The table below shows some of the most common tools used for measuring and marking out on metal.

Tool	Image	Usage
Steel rule		Measuring
Scriber		Marking lines on metal
Engineer's square		Drawing lines at 90° to the edge of the metal or checking for 90° angles
Odd leg callipers		Drawing lines parallel to the edge of the metal
Dividers		Marking circles and arcs on metal

Metal can be cut to size with a hacksaw, junior hacksaw or tin snips, depending on the thickness of the metal. Here are some of the most common metalwork tools used to shape metal.

Tool	Image	Usage
Hacksaw		Sawing straight cuts
Junior hacksaw		Sawing straight cuts through medium to thin pieces of metal
Tin snips		Cutting thin sheets of metal
Raw hide mallet		Shaping sheets, rods or bars of metal without denting the surface
Ball pein hammer		Shaping metal (flat end for flat work and rounded end for curved work)
File		Shaping metal and smoothing the edges

MATERIALS AND MANUFACTURING

Large sheets of metal or strips of metal can be cut using a **guillotine**.

A factory worker feeds a sheet of metal into a guillotine to be cut to size.

Make the Link

Metal is filed in the same way as plastic, using the four stages of smoothing the edges of material.

A smaller version of the guillotine is the **notcher**, which removes little notches of metal to create an outline shape on a sheet or strip of metal. Tin snips can also cut thin sheet metal. The notcher is also known as the 'nibbler' because of the way that it removes metal with a sharp cutting tooth.

After all these processes, the edges of metal can be very sharp. Filing metal transforms rough and ragged edges into smooth edges in the four simple steps, CDEF (see page 151).

Test your knowledge

Cutting and shaping metal

1. State three tools used for marking out on metal.
2. Name a tool for marking lines at 90° to the edge of a metal bar.
3. Name a saw that can be used to create straight cuts on medium to thin pieces of metal.
4. Name a saw that can be used to create straight cuts on medium to thick pieces of metal.
5. Explain one benefit of using a raw hide mallet when working with metal.
6. Describe why the thickness of metal and the angle of the bend must be taken into consideration when bending metal.

To shape metal, it can also be bent or folded. Malleable metals, such as aluminium, can be bent or folded without the need to heat the metal. To achieve a fold or bend in metal without a metal press, a wooden block or **former** can be used. Metal is forced around the former using a raw hide mallet. When bending or folding metal, the thickness of metal and the angle of the bend must be taken into consideration to avoid metal fractures (cracks along the bend line). Bending bars can also be used to shape metal using an engineer's vice.

Metal threads

A '**metal thread**' is the term used when a screw and thread are created with a **tap and die**. To cut an internal thread on metal, a hole must be drilled. The diameter of the hole must be slightly smaller than the diameter of the bar that will fit into it. For example, to fit an 8-mm diameter bar, a 7-mm hole will be drilled. This is known as the **tapping size** as the tool used to cut an internal thread is called a **tap**. Once the hole is cut, three different taps will be used to cut the internal thread:

- **Taper tap** – used first, this tap has sloping sides to make an easy first cut of the thread.
- **Intermediate tap** – has less sloping sides, which cut deeper into the metal.
- **Plug tap** – used last, this straight-sided tap will cleanly cut the metal to the final shape.

The technique for using a tap is to first apply grease to the tap to prevent it becoming stuck inside the hole. Place the tap at the top of the hole, being careful to set it up vertically. Turn the tap wrench clockwise, pressing down to start the cut. For every full turn clockwise, turn the tap back a quarter turn. This clears the waste metal out of the hole. Forgetting to turn back and clear the waste metal can lock the tap in place and forcing it out can snap the tap, leaving half of it in the hole.

To cut an external thread on a bar of metal, a **split circular die** is used. It is held in a **die stock**, which has three **grub screws** on the outside.

- For the first cut, tighten the centre grub screw into the split. This widens the die by increasing the diameter of the hole that cuts the metal, providing a shallow first cut.
- For the second cut, the centre grub screw is loosened and two outer grub screws are slightly tightened, reducing the diameter of the die to cut deeper into the metal.
- Finally, the two outer grub screws are fully tightened to achieve a deep cut, good quality thread.

> **Make the Link**
>
> To drill metal, the same processes can be followed as with drilling wood and plastic (pages 138 and 153).

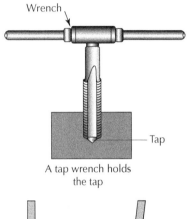

A tap wrench holds the tap

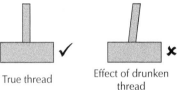

True thread — Effect of drunken thread

Cutting a thread at a wonky angle is difficult to fix and is known as a drunken thread.

> **Hint**
>
> Turn the die in the same way a tap is turned, turning back to clear the waste metal, with grease used to lubricate the cut.

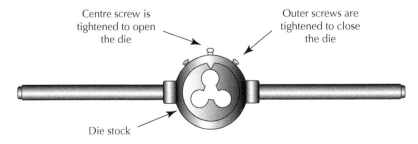

Centre screw is tightened to open the die — Outer screws are tightened to close the die — Die stock

MATERIALS AND MANUFACTURING

Turning metal

The centre lathe is used for shaping metal bars.

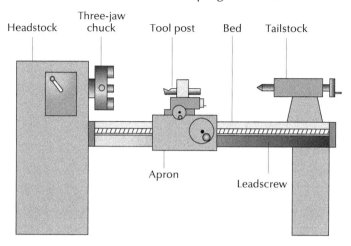

Setting up the centre lathe

The metal blank is secured in the centre in the three-jaw chuck on the headstock using a chuck key. Unlike the wood turning lathe, the tools are not held by hand. A tool post holds the tools and this can be moved in different directions, using the wheels at the front of the machine. A range of indents, angles and patterns can be created with different tools attached on the tool post. For example, the tool post can be set at an angle for taper turning.

Here are some of the most common metal lathe tools.

A chuck key is used to open the three-jaw chuck. Care must be taken to tighten the metal in the exact centre of all three jaws. Clamping the metal between two jaws will mean that metal will not spin centrally and the machine will be unsafe to use. Tighten the three-jaw chuck and slowly open it, trying to slot in the metal. If it doesn't fit, open it a little more. Keep repeating this until it slides in neatly, then tighten it in place.

Remember to remove the chuck key, as, like the pillar drill, this can cause a serious accident if left in.

Tool	Image	Usage
Cutting tools		A range of sharp-edged tools for accurate rounding and shaping
Parting tool		Pointed tool for cutting notches or cutting/removing metal from the lathe
Knurling tool		Creates an indented pattern on the metal

Continued

MANUFACTURING IN THE WORKSHOP

Tool	Image	Usage
Centre bit		Drills a pilot hole in the face of a metal bar
Chuck key		Opens, closes and adjusts the three-jaw chuck that holds metal in the lathe
Micrometer		Accurate measuring diameters

Remember to remove the chuck key. Like the pillar drill, this can cause a serious accident if left in.

There are many different metal-turning processes, as shown in the table below. Each one creates different shapes and profiles.

Tool	Image	Usage
Step turning		Reduces the diameter of the bar but also used to create stepped shapes
Facing off		Smooths the end of the bar
Taper		Cuts metal away at an angle
Knurling		Imprints a texture onto the metal, generally used for extra grip
Drilling		Drills a hole in the face end of the metal
Parting off		Removes the component part from the waste metal
Parallel turning		Reduces the diameter of the bar

Sand casting

Pouring molten metal into a mould to create a solid metal product is called **sand casting**. Sand-cast products are generally big, solid, heavy metal items such as parasol bases, anchors, pipe fittings and machine parts. As the mould cannot be used more than once, it is only suitable for one-off manufacturing.

Drilling on the centre lathe

A centre bit is used for pilot holes on the centre lathe.

The drill bit used first when drilling on the centre lathe is called a centre bit. This creates a small pilot hole. Trying to drill the face of a bar on the centre lathe without a pilot hole will result in the drill bit slipping and scratching the face of the metal. If you are successful in drilling the hole, it is likely that it will not be accurately centred. It is not safe to drill a pilot hole using a twist drill bit as a centre bit is engineered to be far more accurate.

Once the pilot hole is drilled, a twist bit can be attached. The tailstock has a gauge that indicates the depth of the hole.

Safety on the centre lathe

Below are some safety tips for staying safe on the centre lathe:

- Ensure the chuck key is removed before switching on the lathe.
- Ensure the metal bar is secure in the jaw chuck.
- Tuck in loose clothing.

> **Make the Link**
>
> Although the lathe can drill holes in a metal bar, for drilling sheet or strips of metal, a pillar drill can be used. Follow the process for drilling wood on page 138.

✔ Test your knowledge

Centre lathe

1. Describe the four safety precautions to follow while using the centre lathe.
2. Outline the steps to take when securing the metal in the three-jaw chuck.
3. Name the process that reduces the diameter of the bar.
4. Name the process that smooths the end of the bar.
5. State the purpose of the knurling tool.
6. Describe one way in which the tool post can be set up to allow taper turning.
7. State the name of a suitable drill bit for creating a pilot hole on the end face of a metal bar.
8. Describe how to monitor the depth of a hole when drilling on the centre lathe.

Assembly and joining methods for metal

There are many different ways to join metal. Designers select a joining method most suitable for the product they are designing. While there are many non-permanent ways to join metal, such as nuts and bolts, permanent joining methods are a lot stronger. These methods can take more time but they create strong bonds and are extremely durable.

Rivets hold metal together permanently and are often used in large-scale construction, such as buildings and bridges. Smaller versions are used in metalwork.

The Forth Rail bridge was built between 1883 and 1890 using 54 000 tonnes of steel girders joined together by 6 500 000 rivets.

Pop rivets are a quick and easy way to join metal. Special pop rivets (which look like pins) are loaded into a pop-rivet gun (right). Both pieces of metal must be pre-drilled and then the rivet is simply 'popped' through using the gun.

Countersunk, flat-head and round-head rivets are different. They are pushed into a hole, then the back of the rivet is hit with a ball pein hammer and compacted over the hole at the back. This can take a long time, depending on the type of metal the rivet is made from. The join is formed by the rivet as it seals up the hole.

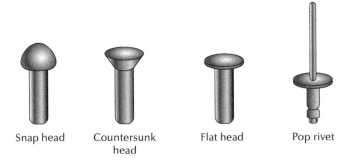

Snap head   Countersunk head   Flat head   Pop rivet

Machine screws, nuts and bolts can be used to non-permanently join metal. These are some of the most common metal fixings and fittings.

MATERIALS AND MANUFACTURING

Fixture/fitting	Image	Usage
Machine screws		Screws that can fit into threaded holes
Nut and bolt		Bolts together metal – a spanner is required
Rivet		Permanently joins metal (see above for the full range of rivets)
Pop rivet		Permanently joins metal – used with a pop-rivet gun
Washer		Used with nuts and bolts to distribute the pressure of the fitting

Welding

Welding is also a permanent method of joining metal and is known as a **thermal joining method** as it uses heat to create the join. There are three types of welding: gas welding, arc welding and spot welding.

Arc welding produces a very bright spark that can seriously damage your eyes, and so a welding mask, with a special black lens, must be worn.

MANUFACTURING IN THE WORKSHOP

- **Gas welding** uses a gas torch to heat a welding rod, which is melted between the two metal parts. When it solidifies, it joins the metal together.
- **Arc welding** works very similarly; however, it uses an electric spark to melt the welding rod.
- **A spot welder** bonds metal together on one specific spot by passing an electric current through the spot. The combination of pressure and electricity creates the join.

Soldering is another thermal method commonly used to permanently join metal.

The three types of soldering are soft soldering, silver soldering and hard soldering (brazing). Each method uses a unique alloy of metal solder wire or rod, which is heated and melts into the join. As the metal cools, it sets and bonds the metal together.

A spot welder uses two electronically charged points to soften and bond metal on one small point, almost like a large staple gun.

Clamping metal

When joining metal together, a clamp is used to hold components in the correct position. The table below shows some of the most common tools and equipment used for clamping, holding and securing metal.

Tool	Image	Usage
Hand vice		Holds small or thin sheets of metal in place while machining or for other general metalwork tasks. Easier to hold than thin sheets of metal
Machine vice		Holds metal in place during machining
Engineer's vice		Workshop vice for general metalwork
Spanner		Tightens and secures nuts and bolts

Test your knowledge

Joining and clamping metal

1. Name a non-permanent method of joining metal.
2. State the name of the tool used when pop riveting.
3. Name the type of hammer to use when joining metal with a round head rivet.
4. Name three types of welding.
5. Describe the difference between arc and gas welding.
6. Describe the process of spot welding.
7. Explain why a hand vice can be used when working with thin sheets of metal.
8. Explain the purpose of vice guards.

Make the Link

The four stages of smoothing the edges of plastic (CDEF) are also used to smooth the edges of metal.

Metal finishes

To prepare metal for a good quality finish, it must first be clean. Rough edges should be filed smooth, using an abrasive paper called emery paper. Like sandpaper, emery paper varies from coarse to fine. Here is some of the equipment for finishing metal.

Tool	Image	Usage
Brushes		Applying paint, varnish, stain or sanding sealer
Steel wool		Smoothing between paint coats
Emery paper		Abrasive paper for smoothing metal
Wire brush		Brushes rust off the surface of metal
Fluidiser		Applying a plastic dip-coated finish

MANUFACTURING IN THE WORKSHOP

Painting metal requires a clean and dry surface. Before applying paint to metal, a primer should be used. Primer is a special undercoat that allows the paint to bond to the metal and enhances the durability of the finish. Applying paint with a brush can make it difficult to achieve even coverage, so aim to build up lots of thin layers. Covering the metal with one thick layer of paint may seem like a quick fix, but paint puddles will form. These puddles take much longer to dry and do not look good. Allow plenty of time to complete one full coat and ensure the paint is properly dry before applying a second coat.

A metal without a finish can be prone to rusting. The function and location of the product must be taken into account when selecting a finish.

Spray paint achieves a much smoother finish. However, the fumes are highly toxic to inhale. Manufacturers spray paint metal in specially designed painting booths, which have extractor fans fitted to remove the toxic fumes.

> **Hint**
>
> Wipe off the excess paint from the brush on the side of the tin or pot to ensure there is not too much paint on the brush.

Plastic dip-coating is a process used to coat metal with a protective layer of plastic. This is ideal for adding extra grip to a metal surface or purely for aesthetic reasons. First, the metal must be thoroughly cleaned so that it is free from dirt and grease, allowing the plastic powder to stick evenly to the surface. Then the metal is heated to around 300 to 400°C. An oven is best for this as the hot air can circulate around the metal to obtain an even heat. The metal is then transferred from the oven to a metal box called a fluidiser. The fluidiser contains a fine plastic powder that is blown around inside. When the hot metal comes into contact with the plastic powder, they bond, creating a plastic coating. This process takes less than five seconds. The plastic will become smooth as it is left to cool.

Plastic dip-coating is a very quick process. However, there are a few common errors that can lead to a poor quality finish. Holding the metal in the fluidiser for too long will build up a thick and unattractive layer of plastic. Also, the plastic powder will not bond to metal that is greasy and so, if the metal is not cleaned properly, the plastic will be patchy. Lastly, the plastic coating will feel gritty if the metal has not been hot enough for it to soften and smooth out.

These dumbbells have been dip coated in plastic.

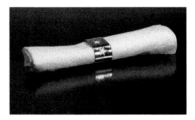

Unfinished metal is quite a common 'finish'. Applying metal polish with a clean dry cloth and lots of effort will help to make the metal shine.

The shiny finish of this metal napkin holder contrasts with the soft, matt texture of the napkin, making the product stand out.

MATERIALS AND MANUFACTURING

Test your knowledge

Link to suggested answers
www.collins.co.uk/pages/scottish-curriculum-free-resources

Plastic dip-coating

1. State two reasons for plastic dip-coating a piece of metal.
2. Explain the reason why the metal must be thoroughly cleaned before it is heated.
3. Describe why an oven is the best way to heat metal before it is put in the fluidiser.
4. Describe three common errors that take place during plastic dip-coating.

Check your progress

I can:

	HELP NEEDED	GETTING THERE	CONFIDENT
describe methods of preparing for manufacture	◯	◯	◯
state the names of appropriate tools, equipment and manufacturing processes for working with wood	◯	◯	◯
state the names of appropriate tools, equipment and manufacturing processes for working with metal	◯	◯	◯
state the names of appropriate tools, equipment and manufacturing processes for working with plastic.	◯	◯	◯

6 Commercial manufacturing

> **By the end of this chapter you should be able to:**
> - describe the benefits of computer aided manufacture (CAM)
> - describe the benefits and drawbacks of computer aided manufacture
> - explain the benefits and drawbacks of 3D printing
> - describe a range of commercial manufacturing processes for wood
> - describe a range of commercial manufacturing processes for plastic
> - describe a range of commercial manufacturing processes for metal
> - explain the impact of commercial manufacturing technologies on the environment
> - explain the impact of commercial manufacturing technologies on society.

An introduction to commercial manufacture

The term 'commercial manufacture' means the manufacturing, assembly and finishing processes that are used to make commercial products or their components. There are four types of commercial manufacture:

- **One-off manufacture** – this is when one product is manufactured at a time and each product is unique. Highly skilled workers are required for this type of manufacturing. In a commercial setting, specialist machinery is used to create **bespoke** high-demand products. Consider a business that sells metal gates that are made to order in a specified size; the products on offer are in high demand, yet each product is unique.

- **Batch manufacture** – a small number of identical products are made at one time. One batch is manufactured before the next batch is made. Highly skilled workers are usually required for this type of production method, but it can also take place in a commercial setting using machines. An example of a product made by batch manufacture is a set of four matching handmade wooden chairs that are manufactured to order.

Make the Link

Consumer demand, as explained in chapter 2, can dictate the type of production method, as the volume of production should match the demand for the product.

MATERIALS AND MANUFACTURING

> **Make the Link**
>
> At National 5 you must understand mass and one-off manufacture, while it is good to have an awareness of continuous flow and batch commercial manufacture, due to their social, economic and environmental impact. Higher Design and Manufacture looks in more depth at the production and planning systems related to these types of manufacture.

- **Mass manufacture** – large volumes of identical products or components are manufactured. The system is highly automated with a small number of workers and, usually, very little manufacturing skill required. An example of this is the manufacturing of kitchen utensils in high volumes using injection moulding machines.

- **Continuous flow manufacture** – extremely high volumes of products or components are produced 24 hours a day, 7 days a week. This system is also highly automated with a small number of workers and very little manufacturing skill. An example of this is the production of plastic water bottles in a factory. The consumer demand for this product is extremely high, therefore the production rate must also be high; one worker may be responsible for a machine that produces tens of thousands of products in one shift.

High speed CAM systems like this water-bottle production line will mould the bottle, fill it with water, cap it and label it by the time it gets to the end of the line.

Computer aided manufacture

In industry, automated mass-manufacturing processes produce a high volume of identical products at a fast pace. As the machinery is controlled by computers, this type of manufacture is named 'computer aided manufacture' (CAM).

The benefits of CAM

The benefits of CAM are:

- high volumes of identical products can be produced
- costs are low when high volumes are produced
- products or components are produced at a fast pace
- a high level of detail is possible
- continual production is possible
- waste is minimised as precise amounts of materials are used.

CAM is extremely quick, compared to traditional methods, and gives high quality products. It can also eliminate human error in manufacture, reducing waste and assuring quality. Products or component parts can be made to an extremely precise level of detail, creating features such as threads or snap-fit joining fittings to the exact size. The set-up costs of CAM are expensive. However, the cost per product is low when thousands or millions of products are produced. This is called **economy of scale** as high volumes of identical products are produced at a fast pace.

COMMERCIAL MANUFACTURING

The drawbacks of CAM

Despite the benefits, there are issues with CAM that can cause production to cease. The drawbacks of CAM are:

- machine breakdowns can cause production to stop completely
- a reduction in employment opportunities in manufacturing industry
- machinery, maintenance and tooling costs are high.

Occasional machine breakdowns interrupt manufacture and production is halted until repairs are made. Manufacturers employ or subcontract maintenance workers to keep machines in continuous production. They also train staff to repair machines and even to spot potential breakdowns before they occur. This incurs a cost to be set against profit.

With a CAM system, small numbers of people are required to manufacture extremely high volumes of products; due to the highly automated systems of CAM there are fewer jobs available for people compared to traditional methods of manufacture. Additionally, there is less demand for skilled workers. While the cost of labour is low, the machinery, the tooling (mould) and maintenance costs are high.

A series of milling machines are controlled by a computer to mass manufacture table legs using CAM technology.

Of course, there are other costs to consider, such as retail, packaging and transportation. When all these costs are considered, CAM only becomes cost effective per product when large volumes are produced.

Employment in new areas in Scotland is emerging to fill the gap in the manufacturing industry, such as the demand for more warehouse staff due to the rise of internet shopping, while in recent years there has been a spike in the popularity of locally produced products.

Highly engineered components can be manufactured as one-piece products while having moving parts within them.

✔ Test your knowledge

CAM
1. State what CAM stands for.
2. Explain three benefits of CAM.
3. Explain three drawbacks of CAM.
4. Describe two ways in which CAM impacts the employment of skilled workers.
5. Explain one way in which the manufacturing industry in Scotland is growing.

MATERIALS AND MANUFACTURING

3D printing with wax instead of plastic allows for the product to be cast using 'lost wax casting' which is a version of casting used in jewellery manufacture.

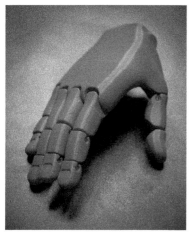

This prosthetic hand made by a 3D printer features moving joints.

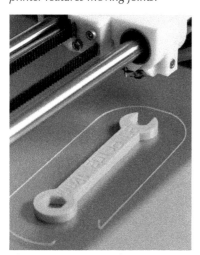

This spanner was made on a 3D printer from ABS plastic. It can be used to test the function of the product before it is mass manufactured.

3D printing

3D printing is a CAD/CAM process that generates a 3D model from a computer-aided drawing. 3D printing machines print one thin slice at a time. Layer upon layer is built up and these layers join together to form a 3D object. Due to its ability to manufacture one product at a time, it is currently used for one-off manufacturing; however, it is possible to create batches of small 3D printed objects at one time. Due to the cost of equipment, it is an expensive manufacturing method if it is not used frequently. 3D printing companies must ensure that they have demand for their products to manufacture enough products to cover the cost of the technology, staff and materials.

3D printing uses

3D printing is ideal for prosthetics as, combined with high-tech body scanning technology, parts can be made to fit a person's unique body shape. 3D printing is an effective way to manufacture interlocking and moving components, as they can be made from one piece. 3D printing is also used during designing to make prototypes so that the client can see the product as a high-quality prototype.

3D printing benefits

The main benefits of 3D printing are:

- it is reasonably fast to produce a prototype compared to traditional manufacturing methods
- the prototype can be made from the actual material that the product will be made from
- the end product can be used in 'rapid tooling' to make injection moulding dies
- the level of accuracy is extremely high
- moving parts can be created
- models can be adapted and reproduced quickly.

3D printing drawbacks

While 3D printing is a well-known method of manufacture, it is an emerging technology. We are yet to utilise it to its full potential in the commercial manufacturing world and research into developing 3D printing is ongoing. As a result:

- the cost of 3D printers is high
- staff require training to use the equipment

COMMERCIAL MANUFACTURING

- technology continues to improve, so ongoing training is essential
- the size of the product is limited to the printer size
- future job losses will take place if 3D printing takes over from manufacturing methods.

> ### ✓ Test your knowledge
>
> **3D printing**
> A hip bone is manufactured by a prosthetics company using a 3D printer.
>
> 1. State five benefits of 3D printing.
> 2. State five drawbacks of 3D printing.
> 3. Describe why the hip bone will be an accurate fit for the user.
> 4. Outline one way in which the prosthetics company can ensure that their 3D printing business is profitable.
> 5. Name five other products that have been 3D printed.

Commercial wood and manufactured-board processes

The wood processes and manufacturing methods explained in Chapter 5 can all be used in commercial manufacture. Many can be combined with automated systems to allow them to take place without a skilled worker. For example, a computer numerically controlled (CNC) drilling machine can be programmed to drill holes in a product as it passes through the production line. In fact, a range of cutting, drilling, shaping and turning processes can be automated on a production line, allowing identical wooden products to be manufactured accurately with consistent quality. It is even possible to apply some types of finish with an automated system.

> ### Make the Link
> Applying lacquer on a mass-manufacturing production line is a way of finishing wood, as explained in Chapter 5.

Commercial plastic processes

A range of cutting, drilling, forming, shaping and finishing processes for plastics can be automated to produce plastic products. In mass manufacture, large-scale plastic moulding methods can produce thousands of identical products at a fast pace. The machines are all highly automated. The four main industrial plastic processes are:

- **laser cutting**
- **injection moulding**
- **vacuum forming**
- **rotational moulding**

> ### Make the Link
> There are many more plastics processes, such as: extrusion, laminating, compression moulding, calendaring and casting. These processes are part of the Higher Design and Manufacture course.

Laser cutting

Designs drawn on specialist CAD software can be sent to the laser cutter, which can cut or engrave on plastic, paper, card or thin sheets of wood or manufactured boards. When cutting plastic, laser cuts cleanly through the plastic, leaving a smooth and even finish; however, some laser lines can be left on the material, which need to be buffed off with wet and dry paper and polished. In commercial factories that can be done with a polishing machine.

The benefits of laser cutting

- The level of accuracy and detail when cutting is very high.
- The laser cutter can also engrave text, numbers and patterns onto the plastic.
- Laser cutters can also calculate the most efficient layout of cutting to reduce waste.

The drawbacks of laser cutting

- The cost of laser cutter technology is high.
- Staff require training to use the equipment.
- Technology continues to improve, therefore ongoing training is essential.
- The size of the product is limited to the laser cutter size.
- Future job losses will take place if laser cutting takes over from manufacturing methods.

Make the Link
A laser cutter can cut through a range of different materials.

More powerful laser cutters can also engrave on or cut through metal.

COMMERCIAL MANUFACTURING

Case study

Bonnie Bling laser cut plastic jewellery

Based on the Isle of Bute, Bonnie Bling is a range of quirky, tongue-in-cheek acrylic jewellery and fashion accessories created by graphic designer Mhairi Mackenzie.

Since the company began, Bonnie Bling has featured in various magazines and website articles and attracted a celebrity audience including Lana Del Rey, Laura Whitmore, Olly Murs, Amelia Lily and even Sir Elton John!

In February 2011, Bonnie Bling scooped the prize for best new product in the jewellery category at Scotland's Trade Show. Since then, the brand has grown from strength to strength. Collaborations with Obscure Couture, RAKSA, Emily Moir and the Riverside Transport Museum have all created unique design-led pieces.

Test your knowledge

Laser cutting

An acrylic clock is manufactured using a laser cutter.

1. State three benefits of laser cutting compared to traditional manufacturing methods.
2. Describe two identifying features of laser cutting.
3. Explain one reason why the clock design is improved with engraving.
4. Name five other products that have been laser cut.

Vacuum forming

This method is commonly used to manufacture packaging as it is suitable for creating complex shapes with thin sheets of plastic. The plastic trays and packaging found in selection boxes, chocolate boxes, biscuit packaging and chocolate advent-calendar trays are vacuum formed. In mass manufacturing, vacuum formers create many items at once and then they are trimmed to size using a guillotine. In one-off production, the waste plastic can be trimmed using a fret saw, coping saw, tin snips or even scissors if the plastic is thin.

Hint

Remember vacuum forming as the process with a sweet tooth.

The process of vacuum forming

Before vacuum forming, a pattern must be made. This is the block the plastic will be formed around and it is usually made from metal or wood. The pattern must have rounded edges to prevent it from tearing the plastic and the sides of the block must be tapered to allow it to be removed easily at the end of the process.

The four stages of vacuum forming are shown below.

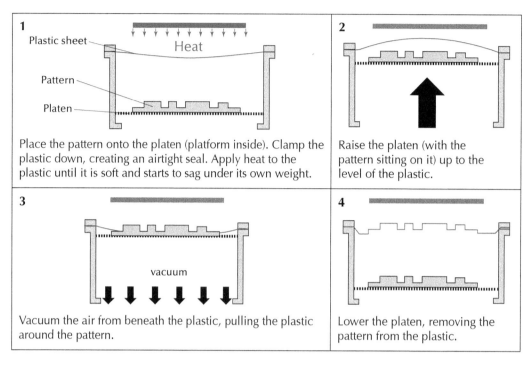

Identifying features:

- Rounded corners
- Tapered edges
- Thin sheet plastic

✓ Test your knowledge

Vacuum Forming
An egg box is manufactured using vacuum forming.

1. State two benefits of vacuum forming.
2. Describe two identifying features of vacuum forming.
3. Explain why vacuum forming is suitable for the egg box.
4. Name five other products that have been vacuum formed.
5. Describe a way in which the egg box could be made more sustainable.

Injection moulding

This method is used to produce plastic products with complex shapes. Plastic granules are heated until molten, then they are forced into a mould where the plastic component or product quickly sets. The mould then opens up and small metal ejector pins force the plastic component or product out of the mould.

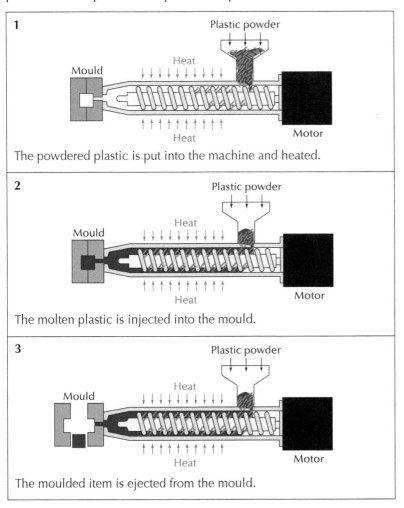

1. The powdered plastic is put into the machine and heated.

2. The molten plastic is injected into the mould.

3. The moulded item is ejected from the mould.

Injection moulded plastic products include toys, remote controls, boxes, buckets, toothpaste caps, chairs, utensils, door stops, make-up packaging, water pistols, phone casings, combs, hairbrushes, car interiors, coffee-cup lids and a whole lot more! Component parts can be made to a high tolerance, meaning that they can snap together due to the accurate sizes. An example of this is a bottle cap that is manufactured so accurately that it secures firmly into place.

MATERIALS AND MANUFACTURING

Identifying features:

- An injection point or sprue mark (a protruding rough mark)
- Ejector marks (round indents)
- Tapered edges
- Complex detail
- Mould split lines (where the mould has joined)
- Webs that strengthen the shape

✔ Test your knowledge

Injection moulding

A toy truck is manufactured from four different plastic components.

1. State two benefits of injection moulding.
2. Describe four identifying features of injection moulding.
3. Explain one reason why the toy truck components can be assembled with no need for any fixings.
4. Explain why injection moulding is suitable for mass manufacturing the toy truck.
5. Name five other products that have been injection moulded.

Rotational moulding

This method is used to produce larger hollow plastic products. Rotational moulded products include children's play equipment (plastic slides and play houses), gritting bins, traffic cones, water tanks, canoes, floats and buoys.

Rotational moulding process

1. Plastic powder is weighed out and placed into the mould.
2. The two halves of the mould are then joined together and sealed tightly; then the mould is heated from the outside, which makes the plastic powder melt.
3. The mould is rotated on a large frame and the plastic powder coats a plastic covering inside the mould. The heat is then switched off and then the plastic will set as it rotates around jets of cool air or cold water.
4. When the two halves of the mould are separated, the hollow product of component can be released from the mould.

COMMERCIAL MANUFACTURING

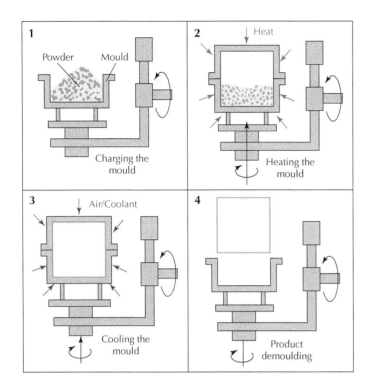

Identifying features:

- Hollow plastic product
- Mould split lines (when the two halves of the mould have joined)

🔍 Case study

One Foot Taller

One Foot Taller is a product design company specialising in the design and supply of lighting and furniture. It began in Glasgow in 1995 and since 2007 it also has a studio in Saint Plancard, France.

The company tries to use local manufacturers (cutting down on transportation) and recycled materials in its super sleek modern designs. They believe in longevity, simplicity and sustainability.

One Foot Taller has dabbled very successfully in interior design, have worked with many other design companies and have an impressive list of products designed for local businesses. It has also designed products for large retailers, such as Marks & Spencer.

In 1999 it designed the Chasm chair (right), which was a rotational moulded hollow form that was then cut down the middle. The two halves were turned around and bolted together, creating two hollow sides. The chair won the Peugeot Design Award, the Blueprint Editor's award at 100% Design (a London design exhibition) and was also awarded Millennium Product Status.

one foot taller

MATERIALS AND MANUFACTURING

✓ Test your knowledge

Rotational moulding
A child's chair is manufactured using rotational moulding.

1. State two benefits of rotational moulding.
2. Describe two identifying features of rotational moulding.
3. Explain why rotational moulding is suitable for the child's chair.
4. Name five other products that have been rotational moulded.

Commercial metal processes

Just as with wood and plastic, a range of metalwork processes that are carried out in the workshop – such as cutting, shearing, notching, turning, drilling, folding and bending – can also be adapted for mass manufacturing. For example, in a factory production line a series of computer numerically controlled (CNC) robotic arms can place and then weld metal components to build a metal car body. Each weld, from one car to the next, will be identical in its quality and location, and the speed of welding will be extremely fast.

A car production line with CNC robotic arms that are programmed to weld the car body.

Most metal manufacturing processes can be carried out by machines. However, sand casting is not suitable for mass production. Instead, the process of die casting is used for casting of metal products to be mass manufactured.

Die casting
In die casting, a piston injects molten metal into a mould, called the die. When the metal solidifies, the two sides of the die separate and ejector pins push the metal product out of the die. Die casting moulds can be used over and over, giving a quick cycle time and large volumes of products.

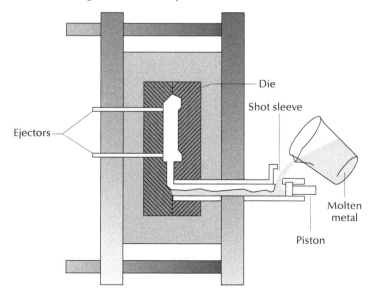

The diagram above shows how one product or component can be made by die casting. However, in commercial manufacture, one die could make numerous casts.

This process can be set up to run automatically and, therefore, high volumes of identical products can be produced at a fast pace. The volume of production, combined with the automation of the system, keeps staff costs down. However, the machinery, the die itself and maintenance costs can be high. Overall, die casting is usually cheap when production volumes are high.

Identifying features:

- Solid metal product
- Intricate details or surface pattern
- Sprue marks (small marks left when the sprue was snapped off)
- Ejector pin marks

Sand casting

Pouring molten metal into a mould to create a solid metal product is called sand casting. Sand cast products are generally big, solid, heavy metal items such as parasol bases, anchors, pipe fittings and machine parts. As the mould cannot be used more than once, it is only suitable for one-off manufacturing.

Sand-casting patterns

The sand-casting process involves first making a **pattern**. This is a full-scale block, cut to the exact shape that is to be cast. For the pattern to release easily from the sand at the end of the process, the sides of the pattern must be tapered.

Sand-casting process

1. A special type of casting sand is sieved to give a smoother finish to the final product.
2. The bottom box, called the drag, is filled with this sand, which contains oil or water to bond it together.
3. The wooden pattern has sand carefully packed around it to make the bottom half of the mould shape.
4. Separately, the top half of the box, the cope, is filled the same way as the drag.
5. The pattern is pressed into the sand to make the top half of the mould shape.
6. The pattern is then removed, the cope is then placed on top of the drag and wooden pins, called *sprue pins*, are used to create cone shaped indents in the sand called the runner and the riser.

The accuracy of die casting is very precise and a high quality finish or patterned surface texture is possible. The text on this rock climbing carabiner has been formed into the metal during die casting.

Many small shapes can be formed at once, joining with sprues that are thin enough to snap off.

● MATERIALS AND MANUFACTURING

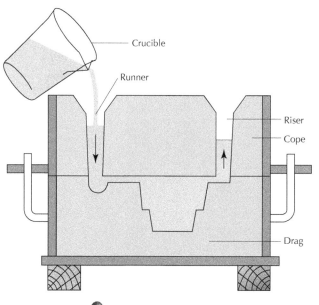

7. Molten metal is poured down the runner and gathers in the mould.
8. Once it is full, the molten metal will flow upwards out of the riser. This indicates that the mould cavity is full.
9. The cast metal requires time to cool and set, then it can be removed from the sand.
10. The product or component requires further finishing as the runner and riser also set as part of the casting.
11. These two parts can be removed with a hacksaw and can be re-used in future casting. The marks left on the metal by the hacksaw are called fettle marks.

Identifying features:

- Solid metal product
- Fettle marks
- A rough surface (caused by the texture of the sand)
- Draft angle (sloping sides)

✔ Test your knowledge

Sand casting
An iron pan is manufactured using sand casting.

1. State two reasons why sand casting is suitable for the iron pan.
2. Describe two identifying features of sand casting.
3. Explain why sand casting is a time-consuming process.
4. Name five other products that have been sand cast.

🔍 Hint
Because in Europe labour costs are high and workers have a very different work/life balance, the Far East remains the prime location for low-cost manufacturing.

Impact of design and manufacturing technologies

Over the past century, there has been a rise in consumerism due to increased global population, and the introduction and availability of affordable products. As a result, design and manufacture has evolved into a global business.

This **globalisation** has brought consumers cheaper products and a wider choice. However, globalisation may also threaten the global economy – such as when an economic crash in one country has a global impact. There are also social issues, such as employment, associated with the movement of manufacturing from one country to another. The environmental issues surrounding globalisation are of huge significance to both designers and manufacturers.

Changes to workforce

With the introduction of manufacturing by highly automated machinery comes a reduction in staff numbers and, consequentially, loss of a skilled workforce. As machines become more complex and are increasingly capable of completing manufacturing tasks to extremely precise detail, the workforce becomes less skilled in manufacturing methods and more skilled in operating computer-controlled machinery. Therefore, the skillset of factory workers has shifted over the past century, while the number of employment contracts has fallen.

> **Hint**
> Importing products also has its drawbacks logistically, such as damage during transportation.

This can lead to:

- the loss of specialist craft skills
- unstable jobs
- increased unemployment
- the closure of factories
- the decline of industrial towns and villages
- economic decline of whole geographical areas.

Additionally, countries in the Far East, like China, Japan and Taiwan, have a different approach to work. The workers are highly efficient with a hard-working attitude. 12-hour shifts are common (compared to an average 8-hour shift in Europe) and, in some factories, workers can work up to 80 hours a week with no days off. Compared to Europe where labour costs are higher and workers have a very different work/life balance, the Far East remains the prime location for low-cost manufacturing.

A modern Chinese electronics factory.

Environmental impact of commercial manufacture

Designers have a responsibility towards the environment, especially when creating a product that will be mass manufactured. During the development of a product, designers must consider the environmental impact of all their design decisions, working with the manufacturer to ensure that the commercial manufacture of the product can be sustainable.

> **Make the Link**
> The philosophy of cradle-to-cradle design, as explained in Chapter 4, considers the impact of the entire product and its potential recycling by-products on the environment.

MATERIALS AND MANUFACTURING

> **Make the Link**
>
> In chapter 2, the product life cycle diagram shows the environmental impact of each stage, from sourcing raw materials to the disposal of the product at the end of its life.

Designers should make recommendations such as:

- using alternative materials that are recyclable
- reducing the volume of material (for example, thinner walls on plastic products)
- reducing the number of parts/components
- designing the product for disassembly
- using recycled components/parts.

The processes and materials used to manufacture, assemble, finish and package the product can pollute the air, water and land. Manufacturers have an environmental responsibility to eliminate pollution and strive towards **clean manufacturing**. This approach involves:

- **n**on-toxic materials
- **o**ptimised raw material use
- **w**ater reductions, to create less waste water
- **a**ir emission reductions
- **s**olid and hazardous waste reductions
- **t**ransport and packaging reduction
- **e**nergy efficiency, and use of solar, tidal, wind or other renewable energy sources.

In 2006, the Dundee Michelin factory was the first Michelin factory in the world to embrace wind energy with two wind turbine generators helping to reduce environmental impact and energy consumption.

Activity

Global manufacturing

This is a group activity that does not require any resources.

1. Discuss the advantages of manufacturing the MP3 player outside the UK.
2. Discuss the drawbacks of manufacturing the MP3 player outside the UK.
3. Report back to the class.

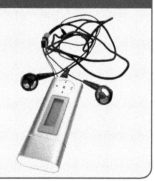

Methods to support sustainability

We are increasingly buying products to improve our lives and the societies we live in. This is called **consumerism**. Often we demand affordable products, sometimes without any consideration of how our decisions impact the environment.

For every product we buy, we should be asking ourselves:

- Who has manufactured the product? What age were they? Were they paid fairly? Were the work conditions acceptable? How long were the working hours? Were the terms of employment fair?

- What product is this one replacing? Can the old product be repaired or maintained? Is the old product recyclable? Is the old product a potential pollutant? Is it really worth replacing?

- Where did the materials come from? Were they locally sourced? Were they transported in an environmentally friendly way? Were they replaced or are they a finite resource? What was the impact on the local environment caused by removing these materials?

- When will the next version or upgrade be available? How long will you use the product? Is it a worthwhile purchase considering its potential lifespan?

- Why do you want or need the product? Is it a necessary purchase? How quickly will it go to waste?

- How do I dispose of the product when I no longer want or need it?

As consumers we have a responsibility to reduce the amount of waste we produce and to reduce the number of products we buy. It is also obvious that we should recycle products and their packaging.

Check your progress

I can:

	HELP NEEDED	GETTING THERE	CONFIDENT
describe the benefits and drawbacks of computer aided manufacture	◯	◯	◯
explain the benefits and drawbacks of 3D printing	◯	◯	◯
describe a range of commercial manufacturing processes for wood	◯	◯	◯
describe a range of commercial manufacturing processes for plastic	◯	◯	◯
describe a range of commercial manufacturing processes for metal	◯	◯	◯
explain the impact of design and manufacturing technologies on the environment	◯	◯	◯
explain the impact of design and manufacturing technologies on society.	◯	◯	◯

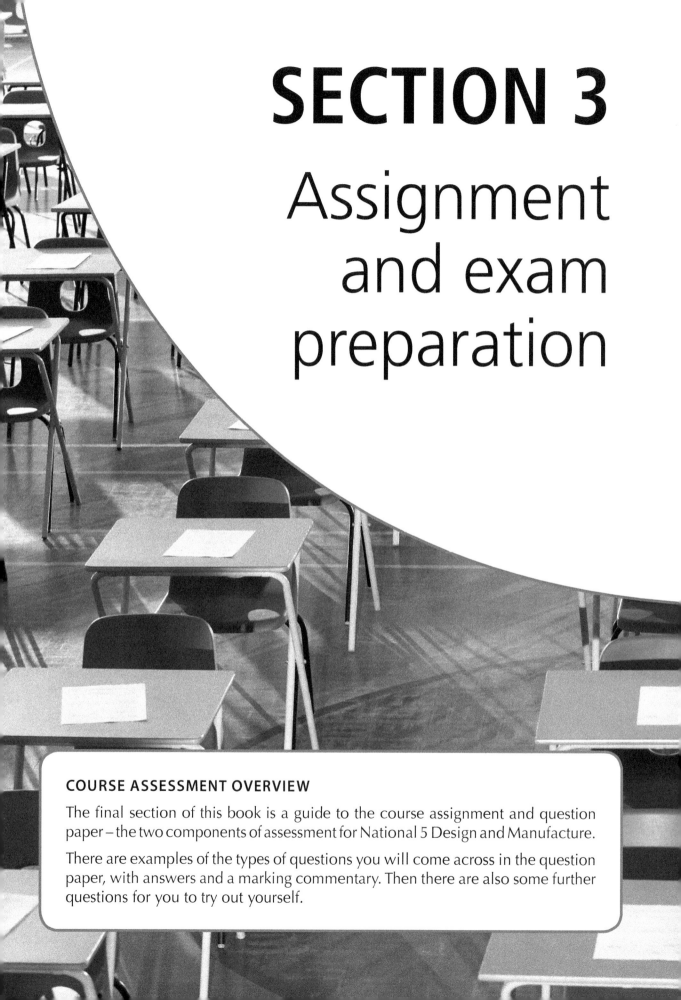

SECTION 3

Assignment and exam preparation

COURSE ASSESSMENT OVERVIEW

The final section of this book is a guide to the course assignment and question paper – the two components of assessment for National 5 Design and Manufacture.

There are examples of the types of questions you will come across in the question paper, with answers and a marking commentary. Then there are also some further questions for you to try out yourself.

7 Assessment

By the end of this chapter you should be able to:
- understand the course assignment
- understand the question paper and how to answer the questions.

National 4 Added Value Unit

The assessment method for National 4 is an assignment called the Added Value Unit, a project that involves designing a product from a design brief given by the SQA, and then manufacturing it in the workshop. This project gives you an opportunity to show your level of knowledge in design, materials and manufacturing methods and then show the level of your practical skills in the workshop.

Grading the Added Value Unit

There is no grade for the added value unit as it is simply a pass or fail that is awarded at National 4. The whole project is marked out of 60 and the pass mark is 42 or above.

Producing a design folio can achieve a maximum of 30 marks:

Task	Potential marks
Carrying out research	3
Adding to a specification	3
Creating a range of ideas	3
Exploring and refining ideas	9
Communicating the development	6
Producing manufacturing details	6
Total for designing	**30**

Manufacturing a prototype in the workshop can achieve a maximum of 24 marks:

Task	Potential marks
Measuring and marking out	3
Using hand and machine tools	9
Assembling components	6
Finishing	6
Total for manufacturing	**24**

Carrying out an evaluation can achieve a maximum of 6 marks:

Task	Potential marks
Evaluating design ideas	3
Evaluating craft skills	3
Total for evaluating	**6**

Added Value Unit evidence

The following evidence is required for the added value unit:

- a design folio — up to six A3-sized pages (including the evaluation)
- a manufactured prototype.

The skills, knowledge and understanding that will be assessed include applying, with guidance, basic knowledge and understanding of:

- research techniques
- idea-generation techniques
- design factors
- graphic techniques
- modelling techniques
- planning techniques
- evaluation techniques
- tools, materials and processes
- manufacturing techniques
- commercial manufacture
- the impact of a range of design and manufacturing technologies on our environment and society.

Preparing for the Added Value Unit

In class, you should have already developed skills and knowledge to be able to complete this unit. As the added value is an assessment, copying work, sharing work or working in teams is not possible. You also cannot ask any other class member for help. Your teacher can give you guidance but will not give you precise details such as what product to design or which material is suitable. They can however point you in the direction of a website or book that contains the information you need, as this is an 'open book' assessment.

To prepare for the added value unit it is a good idea to complete a practice one, such as the exemplar one below.

Exemplar National 4 Added Value Unit

Problem situation

A retail business is looking for new designs for their product range.

They make and sell homeware products and decorative items for the home.

They wish to extend their range for customers with young families, and want to create a product that helps organise items in a hallway, porch or cupboard space.

Design brief

You are required to design and manufacture a working prototype of a product that could organise items in a hallway, porch or cupboard space.

Your task is to produce a design proposal that meets all the requirements of the specification given below and then manufacture a functional prototype.

Specification

The product must:

- be suitable and strong enough to support the weight of other objects
- reflect the style of a young family
- be comfortable to use
- be safe in use.

Task	Evidence	Example of page use
Research and confirm the design brief: 1. carry out research 2. add to the specification	Design work: notes, comments, results of research, annotations, specification	2 pages of A3
Develop design ideas: 1. create a range of ideas 2. explore and refine ideas 3. communicate the development 4. produce manufacturing details	Design work: sketches, models, photographs, annotations, working drawings, dimensioned sketch, cutting list, sequence of operations	3 pages of A3
Evaluating design and craft skills: 1. evaluate design ideas 2. evaluate craft skills	Design work: notes, comments, annotations	1 page of A3

The quality of your prototype is the evidence of your practical ability in the workshop, and combined with your teacher's observation, you must have evidence of:

- measuring and marking out
- using hand and machine tools
- assembling components
- finishing.

National 5 Assessment

There are four key aspects to the Design and Manufacture course: designing, materials, manufacturing in the workshop and commercial manufacture. Throughout the course you will produce a range of evidence that will demonstrate the skills, abilities and knowledge across these areas. The two components that are assessed are the course assignment and the written exam, called the question paper.

Component 1: question paper   80 marks
Component 2: assignment — design   55 marks
Component 3: assignment — practical 45 marks

National 5 Course Assignment

The course assignment is worth 100 marks. A design brief is set by the SQA, not by your teacher. Your task is to design and manufacture a prototype in response to this information. This project should therefore showcase your:

- design knowledge
- design skills
- knowledge of materials
- knowledge of manufacturing
- practical skills.

Course Assignment — Design

The design aspect of the course assignment is worth 55 marks out of a total of 180 marks available for the course assessment. It assesses your design skills by providing you with a set brief for which to design a product. Your design is then manufactured as evidence for the next part of the assignment. You will present this work on a maximum of seven A3-sized sheets or equivalent. A research worksheet and a planning for manufacture worksheet are issued annually with the assignment. Both sides of the research worksheet can be used, but all other sheets must be single sided.

This part of the assignment includes:

- analysing a brief (8 marks)
- generating ideas (9 marks)
- developing ideas (20 marks)
- using models (6 marks)
- using graphics (6 marks)
- planning for manufacture (6 marks)

Course Assignment — Practical

This part of the assignment is worth 45 marks out of a total of 180 marks available for the course assessment. In this assignment your workshop skills will be assessed as you manufacture the design you created in the first part of the assignment. This assignment allows you to design and manufacture a product that reflects your workshop experience, abilities and skill. If you have particularly enjoyed or demonstrated talent with any particular workshop tasks, you should integrate these into your design.

This part of the assignment includes:

- measuring and marking-out (9 marks)
- using hand and machine tools (18 marks)
- assembling components (5 marks)
- finishing (9 marks)
- evaluating (4 marks)

National 5 Question Paper

The question paper is a 80-mark written question paper and is worth 45% of the overall grade. The paper uses command words such as: state, select, outline, identify, describe or explain, and responses are typically written as sentences on lines. However, a few marks per paper may require you to sketch basic diagrams.

There are two parts to the exam, Section 1 and Section 2. You must complete both parts.

Section 1 is worth 60 marks and is mostly based on the materials and manufacturing unit. For question 1, there will be a single extended 30-mark question based on one product, which includes typical workshop processes. There are a further five or six questions in Section 1 that assess design work and aspects of product design. These remaining questions are worth 30 marks.

Section 2 is worth 20 marks. There will be four or five questions in this part of the paper. The first question in Section 2 will assess your knowledge of materials and commercial manufacturing processes. The remaining questions will assess your knowledge and understanding of commercial manufacture and the impact of commercial manufacture on society and the environment.

National 5 Course Specification for the Question Paper

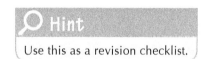

Hint
Use this as a revision checklist.

For the National 5 question paper, you are required to demonstrate knowledge and understanding of the following content.

Section 1, Question 1 content – workshop-manufactured product

Tools for measuring and marking-out:

- the use of measuring and marking-out tools
- callipers: outside and odd-leg
- steel rule
- dividers
- gauges: marking and mortise
- centre punch
- scriber
- squares: try and engineer's

Machine and hand tools for cutting and forming materials:

- the use of hand tools
- saws: coping, tenon, hacksaw and junior hacksaw
- chisels: mortise and bevel-edged
- hammers: ball-pein, cross-pein and claw
- mallets: wooden and hide
- planes: jack, smoothing, rebate and plough
- hand router
- drill bits: twist, forstner, countersink and centre
- files
- pliers
- pop-rivet gun
- screwdrivers
- tin snips
- bending bars
- taps and dies
- nail punch
- bradawl

The use of machine tools:

- sander: disc and belt
- pillar drill: setting-up and depth stop
- scroll/fret saw
- centre lathe: setting-up, parallel and step turning, taper turning, drilling and knurling
- wood lathe: setting-up, preparing material, parting off, parallel turning and finishing
- mortise machine: setting-up and depth stop
- fluidiser
- oven
- strip heater

The use of joining methods:
- adhesives
- screws, nails, nuts and bolts
- woodwork joints: mortise and tenon, lap, rub, halving, dowel, rebate and housings
- pop-riveting
- welding

The use of tools for holding and clamping:
- vices and guards: machine, bench, hand, engineer's
- G clamp
- sash cramps
- the use of formers and jigs

Properties and appropriate use of common materials:
- hardwoods: beech, ash, mahogany and oak
- softwoods: red pine and spruce
- manufactured boards: plywood, flexi-ply, MDF, chipboard and hardboard
- non-ferrous metals/alloys: aluminium, copper and brass
- ferrous metals/alloys: iron, mild steel, high-carbon steel and stainless steel
- thermoplastics: ABS, acrylic, polypropylene and polystyrene thermosetting plastics: urea formaldehyde and melamine formaldehyde

Surface finishing techniques:
- sanding/abrading
- polishing
- varnishing
- oiling
- staining
- waxing
- painting/lacquering
- dip-coating

Section 1, Questions 2–8 content–design work and aspects of product design

Analysis of a brief:
- gathering data
- the key stages of questionnaires and user trips
- reasons for the selection of research techniques
- the role of the product specification in the design process

Idea-generation techniques:
- appropriate use of idea-generation techniques
- the key stages of morphological analysis and brainstorming

The use of modelling in the design process to:
- generate and explore
- test and refine
- communicate
- the advantages of using modelling in the design process

Reasons for selection of types of models:
- sketch
- scale
- block
- computer-generated

The use of graphics in the design process to:
- generate and explore
- test and refine
- communicate
- the advantages of using graphics in the design process
- reasons for the selection of types of graphic techniques

The influence of function on the design of products:
- primary function
- secondary function

The influence of performance on the design of products:
- maintenance issues associated with products
- the influence of a product's life expectancy on design, manufacture and the environment
- fitness-for-purpose of products
- safety issues associated with product

The influence of the target market on the design of products:
- marketing techniques to influence sales
- the benefits of branding
- technology push and market pull

The aesthetics of products:
- influences on the aesthetics of products

The influence of ergonomics on the design of products:
- safety
- comfort
- ease of use
- the use of anthropometric data

Planning for manufacture:
- steps and order
- tools and machines
- safety
- working drawings
- cutting lists

Methods to evaluate products:
- comparison to other products
- user trials
- comparison against specification
- questionnaires

The role of people who influence the design of products:
- designers
- manufacturers
- marketing teams
- consumers
- retailers

Section 2 content–commercial manufacture

Commercial manufacture processes:

- vacuum forming: uses, identifying features and patterns
- sand casting: uses, identifying features and patterns
- injection moulding: uses and identifying features
- rotational moulding: uses and identifying features
- die casting: uses and identifying features
- computer-aided manufacture (CAM): benefits and drawbacks
- laser cutter: uses, benefits and drawbacks
- 3D printer: uses, benefits and drawbacks
- the use of standard components and knock-down fittings
- types of manufacturing systems: mass and one-off

Impact of design and manufacturing technologies:

- on society and the environment
- supply of affordable and accessible products
- changes to workforce
- energy consumption
- pollution
- methods to support sustainability

National 5 exam-style questions

The next pages show a typical question paper, showing the answers written in blue and the marker's commentary in red. The key words and phrases in each answer are in bold.

ASSIGNMENT AND EXAM PREPARATION

Example of a Section 1 question 1 with marking commentary

1. A design proposal for a vanity unit is shown below.

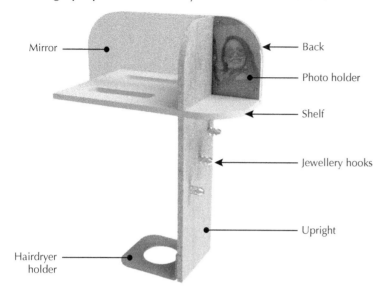

(a) The vanity unit is to be made from different materials.

   (i) Name a suitable softwood for the shelf.   **1**

   Pine

Marking commentary: a mark is awarded for identifying a suitable softwood. Spruce and other softwoods are also acceptable answers.

   (ii) Name a suitable transparent thermoplastic for the photo holder.   **1**

   Acrylic

Marking commentary: a mark is awarded for identifying a suitable thermoplastic.

(b) Templates were used to mark out the MDF back panel, photo holder and mirrored acrylic.

(i) Explain two benefits of using a template to mark out the back panel and mirror. **2**

When marking out curves it is easier to use a template as a compass or dividers may scratch the material. The curves on both parts will match identically.

Marking commentary: marks are awarded for the terms 'quicker', 'easier' and 'faster' if they are explained in comparison to traditional marking-out methods. Creating identical parts is also possible from a template.

(ii) Name the hand tool that would be used to cut the curve on the MDF back panel and the thermoplastic mirror. **1**

Coping saw

Marking commentary: a mark is awarded for the coping saw. Other saws are not accepted as they cannot cut curves or suit both the plastic and the MDF. When a question is worded 'name *the* tool', there is only one possible answer. When a question is worded 'name *a* tool', there is more than one possible answer.

(c) The jewellery hooks will be manufactured on a centre lathe.

The jewellery hooks are manufactured from aluminium.

(i) State two reasons why aluminium is a suitable choice for the jewellery hooks. **2**

*Aluminium is **easy to work with** and cut to shape on the centre lathe. Good **strength-to-weight ratio**, so it is strong enough to be used for hooks.*

Marking commentary: marks are awarded for the properties of aluminium in relation to the needs of the jewellery hook; must be strong enough to hold jewellery. Marks are also awarded for the suitability of using aluminium on the centre lathe.

(ii) Outline two safety checks that must be made to the centre lathe before switching it on. **2**

*Make sure the **chuck key is removed** before switching on the centre lathe. Make sure the metal is **tightly** secured in the **three-jaw chuck**.*

Marking commentary: marks are awarded for the correct checks made on the centre lathe before it is used. These are not checks on the person using the machine, but the machine itself. Answers referring to tucking in loose clothing and tying up hair are checks made to the operator.

[Turn over]

The profile shape below is to be created on the centre lathe.

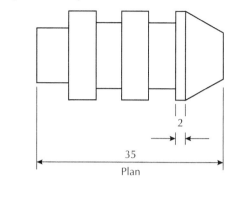
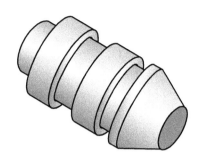
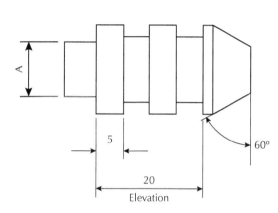
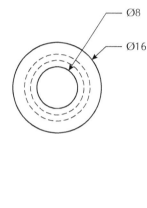

The jewellery hooks are manufactured from aluminium.

(iii) Describe the four processes of turning the metal bar to create the profile shape. **4**

Stage 1 – First, each bar is put into the centre lathe and the end is faced off.
Stage 2 – Change the tool post and create a smoother surface along the metal by parallel turning and make the bar 16mm Ø.
Stage 3 – Angle the tool post to create the 60° tapered edge.
Stage 4 – The parting-off tool is used to cut the grooves and cut the bar to length.

Marking commentary: marks are awarded for identifying the four processes used to create the profile shape. When responding to a question like this where four stages are required to gain four marks, it is a good idea to break down your answer with numbered points (Stage1, Stage 2, etc) to help you form a response that is worth full marks.

(iv) Name a hand tool that could be used to check that diameter A is 10mm. **1**

Outside callipers.

Marking commentary: marks are awarded for naming outside callipers. Following the course specification answers, this is the only acceptable tool listed in National 5 that can read this measurement. However, if the response 'micrometer' was written, it could also gain a mark as technically this is correct.

(d) The shelf shown below was cut and shaped to hold a range of brushes.

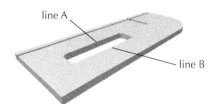

    (i) Name a tool to mark out lines A and B for the slot.   **1**

    Mortise gauge

Marking commentary: a mark is awarded for naming a mortise gauge, as this tool draws two parallel lines. Alternatively, a marking gauge can be named.

    (ii) Describe how the slot could be cut out accurately, with reference to workshop tools.   **2**

    (Sketches may be used to illustrate your answer)

Using a power drill, cut holes at each end of the slot, then use the jigsaw to cut out the waste material.

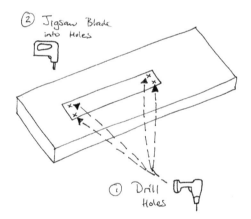

Marking commentary: marks are awarded for a two-step process to remove the waste material. Alternatively, the mortise machine could be used or a coping saw can be used instead of a jigsaw. Additional sketches can help you gain marks if it is difficult to put something into writing. Basic 2D sketches and diagrams with labels like the one shown above can gain marks, so it is worthwhile using the space provided to add a sketch to back up your answer.

[Turn over]

● ASSIGNMENT AND EXAM PREPARATION

The shelf is to be finished with paint.

    (iii) Explain two steps to take when applying the paint to ensure a good-quality finish.   **2**

    *Dip **a brush** in the paint and **wipe off the excess** on the edge of the tin so that a **thin coat is applied**. Paint it on, going in the **direction of the wood grain**.*

Marking commentary: marks are awarded for explaining two aspects of the painting process. In this response, four different points are made: using a brush; not overloading a brush with paint; applying lots of thin coats and paint in the direction of the wood grain. A simpler response may be acceptable, but may not fully explain the steps. In this question, the command word is 'explain', therefore it is important to provide as much detail as you can to make the steps clear.

(e) The hairdryer holder is manufactured from a thermoplastic.

The edges of the thermoplastic were finished before the hairdryer holder was formed.

    (i) Explain why the edges of the thermoplastic should be finished before it is formed into shape.   **1**

    There is less chance of it breaking when it is being held in the vice when it is flat, as it is better supported.

Marking commentary: a mark is awarded for demonstrating an understanding that it is problematic when trying to finish the edges of plastic when it is shaped. Like the template question 1c, a mark is also awarded for the terms 'quicker', 'easier' and 'faster' if they are explained in comparison to finishing the edges when the plastic is shaped. For example, the answer 'it is easier to finish a flat sheet of plastic than a bent one' gains a mark.

Cracks, scratches and edge chips are common problems when working with a thermoplastic.

    (ii) Describe two ways to ensure the thermoplastic doesn't get cracked, scratched or chipped when finishing the edges of plastic.   **2**

    1. *Keep the thin plastic covering on the plastic sheet as long as possible to stop it getting scratched.*

    2. *Keep the plastic low in the vice when filing it to make sure it is fully supported.*

Marking commentary: marks are awarded for identifying workshop methods that prevent cracks, scratches and chips. An explanation is required to gain marks, so a response that simply said 'plastic cover' wouldn't gain a mark.

The thermoplastic has to be heated before forming the 90 degree bend with a jig.

    (iii) State the name of the equipment that should be used to heat the acrylic.   **1**

       Strip heater

Marking commentary: a mark is awarded for identifying the strip heater. In the case of bending plastic along a line, you should always write the correct method of using the strip heater.

Stainless steel screws were used to join the hairdryer holder to the shelf.

    (iv) Describe two benefits of using screws made from a non-ferrous metal.   **2**

       The screws won't rust when they come into contact with water, such as droplets from wet hair.

       Stainless steel is tough, so it will be able to hold the weight of the hairdryer.

Marking commentary: marks are awarded for identifying benefits of a non-ferrous metal, or for more specific properties of stainless steel. Two different benefits must be described to gain two marks.

(f) The upright is to be drilled using a pillar drill.

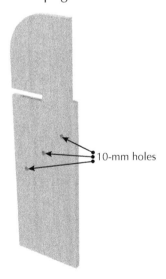

A pilot hole is to be drilled before the 10-mm hole is drilled.

    (i) Explain a reason for drilling a pilot hole.   **1**

       Larger diameter holes drilled without a pilot hole will chip the wood.

Marking commentary: a mark is awarded for explaining the purpose of the pilot hole. Other responses could have included 'to act as a guide for the larger drill bit' or 'prevents the wood from chipping'.

[Turn over]

(ii) Name a drill bit that could be used to drill the pilot holes. **1**

Twist bit

Marking commentary: a mark is awarded for naming the twist bit.

(iii) Name the clamp that should be used to hold the wood while it is being drilled. **1**

G-clamp

Marking commentary: a mark is awarded for naming the G-clamp. As it is seen to be a wide board in the graphic above, a machine vice is not a suitable answer.

(iv) Describe **two** adjustments that should be made to the pillar drill to ensure it drills to the correct depth. **2**

*The maximum depth of the drill can be adjusted by **setting the depth gauge** so that the drill doesn't go the whole way through the wood. Before setting the depth gauge, put the height of the table at a suitable level.*

Marking commentary: marks are awarded for the description of using the depth gauge. This answer does not mention using a steel rule to measure the depth accurately, but describes drilling it to a maximum depth and is therefore worth one mark. Adjusting the table gains a second mark, as the table often has to be moved to put the drill bit in the drill chuck, as someone who has used it before may have adjusted it higher or lower. The drill bit should be close to the work, therefore the table should be adjusted before the depth gauge is set.

Total marks 30

Example of a Section 1 question 2 with marking commentary

2. Designers carry out research to improve the design of products such as the coffee machine below.

Often questionnaires are used as a method of research.

(a) (i) State one piece of information that could be gained from a questionnaire about the coffee machine. **1**

Whether the target market likes the blue colour.

Marking commentary: marks are awarded for statements that show that a questionnaire can gain opinions on product aesthetics, target market information, cost and preferred function, e.g. what beverages do people make in the coffee machine.

(ii) Outline two reasons why a questionnaire is a suitable research technique for the coffee machine. **2**

1. *The questionnaire can be issued electronically and so a wide audience can be reached.*

2. *The results can be analysed and statistics can be made such as '80% have never made hot chocolate with this machine'.*

Marking commentary: marks are awarded for reasons that show that a questionnaire can be collated electronically, creates statistics or other reasons such as a large amount of people being reached in a short amount of time. This question is about the 'suitability' of a questionnaire for the coffee machine and is looking for a different answer than the one given for 2(a)(i) above, which is more specifically looking for the 'information gained' from a questionnaire.

To gain first-hand knowledge of the coffee machine, the designer carried out a user trip.

(iii) Describe three key stages of the user trip. **3**

1. <u>Preparation:</u> *plan out how to use the coffee machine by looking at the instructions and thinking about the ways in which the user will work the machine.*

2. <u>Carry out the user trip:</u> *plug it in, heat it up, put in the coffee and make a coffee while noting things like the ergonomic or functional issues that you find, and take photos of these as evidence.*

3. <u>Collate:</u> *show all the findings of the user trip by putting a collection of photos together and writing up all the findings from the user trip.*

Marking commentary: marks are awarded for a description of the three key stages of a user trip; to prepare for the user trip, carry out the user trip and collate evidence of the activity. To gain full marks for this question, you must refer to all three of these stages. It is a good idea to make reference to the coffee machine, as it helps to structure the answer and you can give examples of what the coffee machine user trip may look like.

Total marks 6

● ASSIGNMENT AND EXAM PREPARATION

Example of a Section 2 question with marking commentary

1. Computer Aided Manufacture is widely used by companies to mass produce products like the disposable coffee cup in high volumes.

(a) (i) Describe two drawbacks of CAM for the workforce in mass production. 2

Fewer jobs are available in the manufacturing industry as machines replace skilled people.

It is cheaper to manufacture in the Far East, so many people have lost their jobs due to globalisation of the marketplace.

Marking commentary: marks are awarded for the correct identification of two ways in which CAM negatively impacts the workforce. The response correctly identifies there being fewer jobs available to those with traditional manufacturing skills, and that labour is cheaper overseas, therefore there are fewer job opportunities. Further answers could include lower salaries due to less-skilled employment, or the loss of specialised skills.

(ii) Describe two drawbacks of CAM in mass production to the manufacturer. 2

The cost to fix machines can be high as not only do machines need to be repaired, but manufacturing halts, which impacts on company profits. Also, the staff need to be trained so that they know how to work the machinery, which again costs the manufacturer money.

Marking commentary: marks are awarded for the correct identification of two ways in which CAM has a negative impact on the manufacturer. The response correctly identifies repair costs and the additional cost of staff training; however, another valid answer is the time taken away from manufacturing for these repairs/training to take place. The response gained the full two marks.

(iii) Describe four ways in which CAM impacts the environment. **4**

The set-up costs of CAM are high and so large volumes of product have to be made to recoup this money. The factories therefore create a lot of air pollution, even 24/7. With this constant production comes constant waste products/materials. Sometimes these are re-used by environmentally aware companies, such as burning wood to heat the factory. When machines break, the old broken parts are also a waste product. The idea that we use 'disposable' products like the coffee cup as an everyday product is not a sustainable way of thinking. People need to consider their own re-usable cup instead of using throw-away products that take decades to break down in landfill.

Marking commentary: marks are awarded for the description of continual air pollution, waste materials, green companies re-using waste material and the need to rethink how we use products as a society. The point made about recycling broken parts is very similar to the point about waste material, so it is always good to ensure that a breadth of reasons is given to achieve full marks. Referring to the product shown will again help you to construct your answer. If no product is shown or perhaps a generic photo of a factory is shown, then you can provide your own product examples to help construct your answer. Other answers could include sustainability issues relating to increased production causing climate change, packaging waste and for a reduction of materials used.

Practice Exam-Style Questions

Now that you have read over the examples, it is time to attempt some exam-style questions on your own.

The following questions are exam-style questions that you may work through yourself.

SECTION 1 – 60 marks
Attempt ALL questions

1. A design proposal for a pull-along toy is shown below.

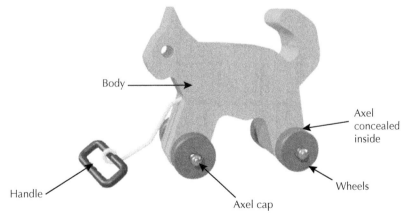

(a) The body and wheels of the pull-along toy are made from wood.

  (i) Name a suitable softwood for the body. **1**

[Turn over]

(ii) Name a suitable hardwood for the wheels. **1**

___

The four wheels were turned from one blank on the wood lathe.

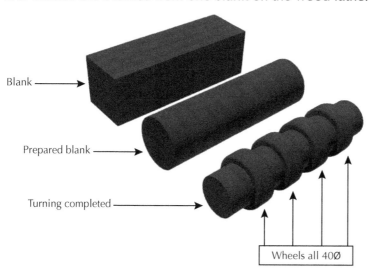

(iii) Describe how to mark out and remove the corners of the blank in preparation for turning. You must refer to workshop tools in your answer. **4**

(Sketches may be used to illustrate your answer.)

___

(iv) Name a hand tool that could be used to check the diameters in millimetres of all four of the wheels to ensure that they are equal. **1**

(b) The wooden block for the body was marked out as shown below and then the holes were drilled.

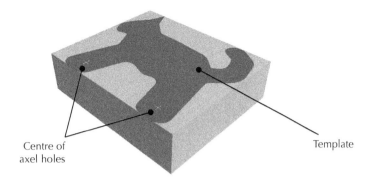

Centre of axel holes

Template

(i) State a benefit of using a template to mark out the cat shape. **1**

(ii) Describe how to mark out the centre points for the axels in preparation for drilling. **2**

You must refer to workshop tools in your answer.

A pillar drill was used to drill the wooden block before it was cut into the cat shape using a scroll saw.

(iii) Name a suitable drill bit to drill the holes for the axels. **1**

(iv) Outline two safety checks that must be carried out on the scroll saw when cutting the wooden block to shape. **2**

[Turn over]

Orange wood stain was applied to the body of the pull-along toy.

(v) Describe three stages of applying wood stain. **3**

(c) An external thread was cut at both ends of the axels.

(i) Name a suitable non-ferrous metal for the axels. **1**

(ii) State a reason for the chamfer at the ends of the axels. **1**

(iii) Name the hand tool that would be used to cut the external thread. **1**

(iv) Describe the process of cutting the external thread on the axel. **3**

(v) Describe one method of checking the thread is a tight fit. **1**

(d) The acrylic rod for the handle was shaped with curved indents before being heated in an oven and formed to shape.

(i) State two reasons why acrylic is a suitable choice for the handle. **2**

(ii) Name the tool used to create the curved indents. **1**

(iii) Name a vice that should be used to hold the acrylic when creating the curved indents. **1**

(iv) Explain why the curved indents should be created before the handle is formed into shape. **2**

The acrylic has to be heated before forming it into shape.

(v) State the name of the equipment that should be used to heat the thermoplastic to the required shape. **1**

[Turn over]

2. Designers researched ergonomics when designing the chair shown below.

   (a) Name two research techniques used to gather research for the chair. 2

   _____

   _____

   (b) Describe three key stages of research when designing. 3

   _____

   _____

   _____

   _____

   _____

   _____

3. A specification is written when developing a design proposal.

   Describe **two** ways in which a specification could be used in the design process. 2

   _____

   _____

   _____

   _____

4. Designers use idea-generation techniques when designing to inspire them to create a range of ideas.

   (a) Name two idea-generation techniques. 2

   _____

   _____

(b) Explain one reason why planning is important before an idea-generation technique begins. **1**

5. A range of sketching, drawing and modelling techniques are used by designers to develop a design proposal.

    (a) Explain why sketch models are a suitable modelling technique for initial ideas. **2**

    (b) Explain two benefits of using 3D drawings, such as isometric drawings, when developing a design proposal. **2**

    (c) Explain two benefits of working drawings when planning for manufacture. **2**

    (d) Outline two benefits of using a block model when communicating a design proposal to a client. **2**

[Turn over]

6. A range of design factors have been taken into consideration when designing the child's chair below.

(a) Describe how ergonomics has influenced the design of the child's chair.   4

A range of marketing strategies are used to promote the child's chair.

(b) Describe two benefits of launching a product under a successful brand name.   2

(c) Outline two other marketing techniques that could be used to influence sales of the child's chair. **2**

7. The coffee pod machine is an example of a technology push product.

(a) Explain the term 'technology push'. **2**

(b) Outline two reasons other than function that would make the coffee machine appealing to the target market. **2**

SECTION 2 – 20 marks

Attempt ALL questions

8. The tool box shown below is made from a variety of different materials and has been mass manufactured.

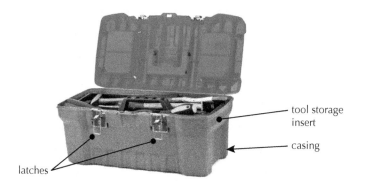

(a) (i) State the name of a suitable material for the tool box casing and explain why it is suitable for this part. **2**

_____

_____

(ii) State the name of a suitable material for the latches and explain why it is suitable for this part. **2**

_____

_____

The tool box casing was manufactured by injection moulding.

(b) Identifying features of this process are visible on the tool box.

(i) Explain the reasons for injection and ejection marks. **2**

_____

_____

_____

(ii) Describe two reasons for including webs on the inside of the tool box. **2**

_____

_____

_____

The latches shown below have been mass produced by die casting.

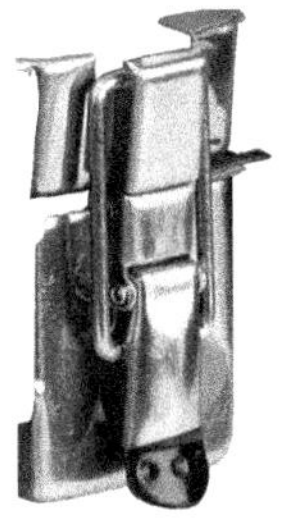

(c) State two identifying features of die casting. **2**

_____

_____

9. Consumers have access to a range of affordable products like this lamp due to the rise in computer aided manufacture (CAM).

Explain two benefits of using CAM to produce affordable products. **2**

_____

_____

_____

_____

_____

[Turn over]

ASSIGNMENT AND EXAM PREPARATION

10. Standard components are used in mass-produced flat-pack furniture.

  (a) Describe two benefits to the manufacturer when using standard components to mass produce products. 2

  _____

  _____

  _____

  _____

  (b) Describe two benefits to the consumer when using standard components to mass produce products. 2

  _____

  _____

  _____

  _____

11. Sourcing materials from all over the world to make mass-produced products has an environmental impact.

Describe two ways in which sourcing materials can negatively impact the environment. **2**

12. Design and manufacturing technologies used in mass manufacture of commercial products have had an impact on the workforce.

Describe the changes to the workforce caused by such technologies. **2**

[Turn over]

ASSIGNMENT AND EXAM PREPARATION

Check your progress

I can:

	HELP NEEDED	GETTING THERE	CONFIDENT
• understand the assignment and how to use the course assignment checklists	○	○	○
• understand the question paper and how to answer the questions.	○	○	○